SpringerBriefs in Physics

SpringerBriefs in Physics are a series of slim high-quality publications encompassing the entire spectrum of physics. Manuscripts for SpringerBriefs in Physics will be evaluated by Springer and by members of the Editorial Board. Proposals and other communication should be sent to your Publishing Editors at Springer.

Featuring compact volumes of 50 to 125 pages (approximately 20,000–45,000 words), Briefs are shorter than a conventional book but longer than a journal article. Thus, Briefs serve as timely, concise tools for students, researchers, and professionals.

Typical texts for publication might include:

- A snapshot review of the current state of a hot or emerging field
- A concise introduction to core concepts that students must understand in order to make independent contributions
- An extended research report giving more details and discussion than is possible in a conventional journal article
- A manual describing underlying principles and best practices for an experimental technique
- An essay exploring new ideas within physics, related philosophical issues, or broader topics such as science and society

Briefs allow authors to present their ideas and readers to absorb them with minimal time investment. Briefs will be published as part of Springer's eBook collection, with millions of users worldwide. In addition, they will be available, just like other books, for individual print and electronic purchase. Briefs are characterized by fast, global electronic dissemination, straightforward publishing agreements, easy-to-use manuscript preparation and formatting guidelines, and expedited production schedules. We aim for publication 8–12 weeks after acceptance.

More information about this series at http://www.springer.com/series/8902

Umut Gürsoy

Holography
and Magnetically Induced
Phenomena in QCD

 Springer

Umut Gürsoy
Institute for Theoretical Physics
Utrecht University
Utrecht, The Netherlands

ISSN 2191-5423   ISSN 2191-5431   (electronic)
SpringerBriefs in Physics
ISBN 978-3-030-79598-6   ISBN 978-3-030-79599-3   (eBook)
https://doi.org/10.1007/978-3-030-79599-3

This Springer imprint is published by the registered company Springer Nature Switzerland AG
The registered company address is: Gewerbestrasse 11, 6330 Cham, Switzerland

Dedicated to the memory of H. Ferda Gürsoy, whose grace, percipience and intellect enriched the lives of all who were fortunate to know her.

Preface

We review recent developments in studies of the strong nuclear force, quantum chromodynamics (QCD), focusing on the strongly interacting regime. We investigate the theory using the non-perturbative technique of gauge-gravity duality, or *holography* in short. After a brief introduction to the salient features of QCD, its electromagnetic properties, ongoing and future planned experiments, and two of the non-perturbative tools, relativistic hydrodynamics and the AdS/CFT correspondence, we present a "bottom-up" holographic model for large N QCD coupled to electromagnetism in detail. We review how to obtain the hadron spectra, structure of the ground state, thermodynamics and energy-momentum transport with a special focus on the electromagnetic probes and their realization in the holographic description.

This monograph has grown out of a series of work on holographic techniques applied to strong nuclear force, in particular, the improved holographic models and their various extensions. The latter was developed in collaboration with Elias Kiritsis, Francesco Nitti, Liuba Mazzanti and Matti Jarvinen. In addition, I am grateful for the collaboration and useful discussions with Francesco Bigazzi, Alex Buchel, Aldo Cotrone, Tuna Demircik, Tara Drwenski, Ioannis Iatrakis, Aron Jansen, Karl Landsteiner, Georgios Michalogiorgakis, Govert Nijs, Carlos Nunez, Andy O'Bannon, Marco Panero, Ioannis Papadimitriou, Giuseppe Policastro, Andreas Schafer and Wilke van der Schee. Subsequent developments in the subject, in the context of V-QCD (the Veneziano limit of improved holographic QCD), resulted from the various works by Timo Alho, Daniel Arean, Francesco Bigazzi, Roberto Casero, Aldo Cotrone, Ioannis Iatrakis, Matti Jarvinen, Keijo Kajantie, Elias Kiritsis, Carlos Nunez, Angel Paredes, Cobi Sonnenschein and Kimmo Tuominen.

Utrecht, The Netherlands   Umut Gürsoy
April 2021

Contents

1 **Introduction** 1
   1.1 QCD and Experiment 1
   1.2 Hydrodynamics 3
   1.3 Holography 4
   1.4 Magneto-Holography 6
   References 7

2 **Basic Features of QCD** 9
   2.1 Lagrangian 9
      2.1.1 Asymptotic Freedom 10
   2.2 Confinement 10
   2.3 Chiral Symmetry and Axial Anomaly 12
   2.4 QCD Thermodynamics 13
   References 14

3 **Hydrodynamics and Transport Coefficients** 15
   3.1 Generalities 15
   3.2 Transport Coefficients 17
   References 19

4 **Gauge-Gravity Duality** 21
   4.1 Holographic Computation of Correlators 22
   4.2 An Example: Holographic Computation of Shear Viscosity 25
   4.3 Holographic QCD 28
      4.3.1 What Do We Want from Holography? 28
      4.3.2 Top-Down Versus Bottom-Up 29
   References 33

5 **Improved Holographic QCD** 37
   5.1 Construction of the Theory 37
   5.2 UV Asymptotics 38
   5.3 IR Asymptotics 43
   5.4 Parameters and the Dilaton Potential 46
   5.5 The Glueball Spectra 48

References 51

**6   Thermodynamics and the Confinement/Deconfinement
      Transition** 53
      6.1   Black-Brane Solution 54
      6.2   Entropy, Gluon Condensate and Conformal Anomaly 57
      6.3   Deconfinement Transition 58
      References 63

7   Improved Holographic QCD at Finite Density 65
      References 68

8   Improved Holographic QCD with Magnetic Field 71
      8.1   Background with Finite Magnetic Field and Temperature 72
      8.2   Quark-Antiquark Potential in a Magnetized State 74
      8.3   Phase Diagram of ihQCD with Magnetic Field 75
      8.4   Inverse Magnetic Catalysis 77
      8.5   Speed of Sound 81
      References 82

9   Hydrodynamics and Transport Coefficients 85
      9.1   Shear and Bulk Viscosity 85
      9.2   Shear Viscosity in the Presence of Magnetic Field 86
      9.3   Anomalous Transport 88
      References 94

10   Conclusion and a Look Ahead 97
      References 99

Appendices 101

Acronyms

QFT   Quantum field theory
QCD   Quantum chromodynamics
ihQCD   Improved holographic QCD
VQCD   Veneziano-QCD
χSB Chiral symmetry breaking
MC   Magnetic catalysis
IMC   Inverse magnetic catalysis

Chapter 1
Introduction

1.1 QCD and Experiment

Quantum field theory of the strong nuclear force, Quantum-Chromodynamics (QCD for short), is one of the most active research fields in high energy physics today both for the theoretical challenges it presents, and for the ongoing and future planned large-scale experimental programs that explore its extreme limits: high temperature and quark density.

The ongoing heavy ion experiments at RHIC (Brookhaven) and LHC (CERN) collide gold and lead nuclei and produce an exotic state of matter called the quark-gluon plasma. This plasma behaves as an almost perfect fluid with the smallest shear viscosity per entropy ratio observed in nature.[1] The fact that this collection of quarks and gluons behave as a fluid instead of a weakly interacting gas of particles is an indirect sign of strong correlations among the constituents, quarks and gluons in this system. There are future planned experiments, NICA (Dubna) and FAIR (Germany) to mention two, which will further explore different limits of QCD again by colliding heavy ions. In particular these experiments will provide further information on the finite quark density regime of QCD. In addition, RHIC isobar experiment has finished its run in 2019 and analyses of the collected data is ongoing at the time of writing this manuscript. This data is expected to reveal crucial information on how the external magnetic fields affect energy-momentum and charge transport properties in the plasma. Another, fascinating new window into quark-gluon physics has recently opened by precision measurements of neutron stars, extremely dense astronomical objects that pack as much mass as two suns within a sphere of radius of about 10 km. The gravitational wave detectors LIGO and Virgo and the X-ray telescope NICER provide crucial new information on QCD at finite temperature and density. The separate regions of the phase diagram that these experiments explore are shown in Fig. 1.1.

---

[1] Barring superfluids where viscosity vanishes.

© The Author(s), under exclusive license to Springer Nature Switzerland AG 2021
U. Gürsoy, *Holography and Magnetically Induced Phenomena in QCD*,
SpringerBriefs in Physics,
https://doi.org/10.1007/978-3-030-79599-3_1

Fig. 1.1 Ongoing (solid) and future-planned (dashed) experiments that explore different parts of the phase diagram of QCD. Inclusion of a third axis, magnetic field, in the phase diagram is believed to enrich this picture substantially

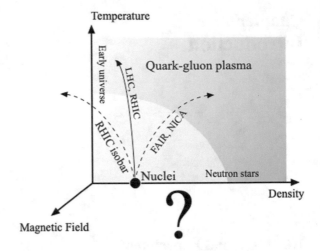

It is believed that the effective QCD coupling constant in the regimes probed by these experiments are sufficiently strong leading strong correlations between particles. As a result, perturbative approximation to QCD, which requires small coupling constant, breaks down and non-perturbative approaches become inevitable. The first-principles non-perturbative approach to QCD is lattice quantum field theory. Placing the theory on a 4 dimensional discretized Euclidean space-time lattice with finite lattice spacing, lattice QCD computes the Euclidean generating function of QCD, from which Euclidean correlation functions of observables follow by taking the continuum limit. One such fundamental observable is the QCD equation of state (EoS), that is pressure as a function of energy or temperature in a thermal state. Thermal states are defined by imposing periodic (anti-periodic) boundary conditions for the bosons (fermions) on the Euclidean time circle whose inverse length defines the temperature of the state. Equation of state is of fundamental importance and a crucial input in analysing experimental results. Lattice QCD is well established to compute the hadron spectrum, the EoS (at vanishing quark chemical potential) and many more fundamental properties of QCD. This, however leaves room to examine other certain observables and regimes which proved hard to study on the lattice. One such regime is finite quark density where the lattice approach meets obstacles due to the long-standing sign problem [1]. Another set of observables, for which the lattice approach becomes difficult, are the real-time correlation functions. The reason is that lattice QCD is inherently defined using statistical averages. This is a natural habitat for the Euclidean correlators and the real-time observables should be defined from these by analytic continuation in time, an operation that generates large systematic uncertainties. It should be noted, however, despite all these difficulties, lattice QCD remains to be the basic rigorous first-principles approach to QCD and the aforementioned problems could be solved in the near future, see for example [2].

All this shows that theory lags behind experiment, and alternative non-perturbative approaches to QCD would be beneficial even if they are approximate. Among the effective approaches the following two somewhat complementary theories stand out for their universality and versatility: *hydrodynamics* an *holography*.

Relativistic hydrodynamics—the conventional theory of relativistic fluids [3]— has been reformulated in the last decades as an effective approach to describe dynamics of "slow variables" in quantum field theory at finite temperature. Assuming presence of thermal equilibrium, one can divide the operators of the theory in two separate classes depending on their behavior under a small, space-time dependent perturbation around this thermal state. If the operator is not conserved, its value can vary at any space-time point allowing it to relax to its equilibrium expectation value quickly. These are "fast variables". On the contrary, conserved operators, such as the energy-momentum tensor, charge current and angular momentum can only change their expectation value at a given space-time point through transport, therefore they relax much slower. Therefore the late time, large wavelength properties of the system is governed by the dynamics of the slow variables. This theory is called hydrodynamics and it is organized in terms of a derivative expansion that corresponds to treating the parameters $\tau_{relax}\ \omega$ and $\ell_{mfp}\ |\mathbf{k}|$ small where τ_{relax}, ℓ_{mfp}, ω and $|\mathbf{k}|$ are the relaxation time, mean free path and the frequency and momentum of a typical dynamical process, respectively.

1.2 Hydrodynamics

Relativistic hydrodynamics provides a good description of transport properties of the quark-gluon plasma as we explain in more detail in Chap. 3. Today, it is generally accepted as a standard tool to analyze energy and charge flow in this plasma [5–10]. This is because particle yields obtained from hydrodynamics simulations match well (except for ultra peripheral collisions) the particle distributions observed in experiments.

However, hydrodynamics is, in some sense, an incomplete theory. It consists of two components: the equations of motion for the conserved quantities, and, the constitutive relations that supplement these equations and express these conserved quantities in terms of hydrodynamic variables such as the fluid velocity in a derivative expansion. Each term in this derivative expansion comes with a *transport coefficient* that is a function of the variables that characterize the thermal state e.g. temperature T, chemical potential μ, magnetic field B etc. The transport coefficients e.g. viscosity and conductivity themselves determine how energy, charge etc. flows in the fluid.

Hydrodynamics does not specify these transport coefficients. Rather, they are determined from the microscopic description of the underlying theory by the Kubo formulas [4]. For example, to determine the conductivity of the quark-gluon plasma, one needs to compute a retarded two point function of the current operator in QCD in the limit of vanishing 4-momentum; to determine the shear viscosity one needs to compute a retarded two-point function of the stress tensor etc. These computations become inaccessible at large coupling constants as the perturbative approach to QCD breaks down. As explained above, lattice approach also has limited applicability for the real-time physics. Today, arguably the most promising tool to determined these coefficients theoretically is the *gauge-gravity correspondence*, or *holography* for short.

1.3 Holography

Holography is a string theory method which, in principle, goes beyond hydrody-namics and determines the thermodynamic potentials and transport coefficients of the fluid [11–13]. The basic idea is to treat the renormalization group energy scale Q of quantum field theory as a 5th dimension and reformulate the effective action of the QFT (in the Wilsonian sense), a function of 4D space-time and Q, as a 5D gravitational theory. One then finds the RG flow, for example how beta functions in the QFT depend on the energy scale, by solving the equations of motion of the 5D gravity theory. This provides the quantum state of the theory. A thermal state with temperature T is given by a 5D black hole solution with Hawking temperature T (see Fig. 1.2). Fluctuations on top of this background determines the correlation functions, from which the spectrum of the theory, the transport coefficients of the fluid state etc. can be read off.

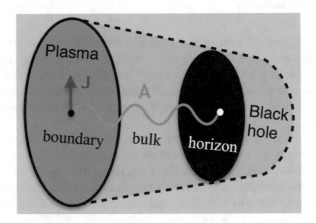

Fig. 1.2 Holographic correspondence posits an equivalence between string theory—which approx-imates to Einstein's general relativity coupled to matter fields at low energies—in the bulk of a 5 dimensional hypothetical space with a black hole at the center and quantum field theory of a fluid on the boundary of this space [11–13]. Originally formulated in IIB string theory on 5 dimensional Anti-de-Sitter spacetime and maximally supersymmetric Yang–Mills theory [11], this equivalence has, since then, been applied to realistic systems, such as the quark-gluon plasma. Concretely, it maps collective flow in the plasma—here denoted by current J at the black point—to fluctuations near the horizon—at the white point—through propagation of bulk wave, A, toward black hole. Universal properties of horizon geometry then lead to constraints on transport in the plasma

However, holography is still a conjecture, not a first-principles theory. It is almost completely established in original example due to Maldacena, e.g. the correspondence between type IIB string theory on the 10 dimensional $AdS_5 \times S^5$ geometry with N units of five-form flux on S^5 and the maximally supersymmetric $SU(N)$ Yang–Mills theory in 4D. In this case, in the limit of large N and strong interactions the correspondence has passed impressively many non-trivial tests which leaves little doubt on its validity. Examples of the correspondence obtained by deforming the maximally supersymmetric Yang–Mills theory by relevant or marginal operators, or the correspondence for different quantum states e.g. thermal, non-equilibrium, finite chemical potential etc. are also generally accepted.

However, QCD is not in these categories and it requires a different holographic approach. Here, the most fruitful route proves to be the "bottom-up" holographic approach that is based on identifying a semi-closed subset of operators relevant for the IR limit—that is in complete analogy with the "slow variables" of hydrodynamics explained above—and constructing the 5D corresponding background by demanding physical requirements. Typically the corresponding 5D background contains a metric corresponding to the stress-tensor, a number of vector fields corresponding to conserved currents, and a number of scalar fields corresponding to scalar operators with non-trivial beta-functions.

This bottom-up approach will be detailed in Chap. 5 below. Here, I will just present a simple handwaving argument in favor of it: just like in hydrodynamics the fact that the IR limit of theory is governed by a finite subset of operators is dictated by the Ward identities. For simplicity let us ignore the fermions and consider pure SU(N) Yang–Mills theory in the large N limit. There are two Ward identities for this theory (1) which conservation of the stress-tensor and (2) which dictates that the trace of the stress tensor being proportional to a single scalar operator, that is the "scalar glueball operator", square of the Yang–Mills field strength. This motivates one to construct a holographic dual in 5D which involves metric and a single scalar field. All other (infinitely many) operators of the pure Yang–Mills theory have vanishing beta functions. The effect of possibly non-trivial vacuum expectation values of these fields are assumed to be implicitly contained in the 5D background geometry.

Even though this is not a first principles approach, and necessarily involves uncontrolled approximations, it is based on reasonable assumptions and quite attractive for its simplicity. It has been shown that it reproduces thermodynamic properties of QCD from lattice to a very good accuracy and extends these results to finite chemical potential and real-time phenomena. From a more general point of view, even a generic tool for qualitative "universal" results—for example specific relations between its transport properties—provides valuable insights in the absence of any other non-perturbative tools.

Fig. 1.3 Large magnetic fields are generated in off-central heavy ion collisions. These fields affect the charge dynamics in the quark-gluon plasma created in these collisions providing an important probe into electromagnetic properties of QCD. Figure taken from [14] and reprinted with permission from Elsevier publishers

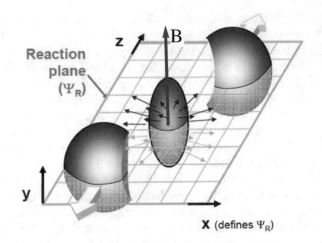

1.4 Magneto-Holography

This review also focuses on the recent developments in the field of electromagnetic properties of strongly interacting gauge theories, in particular QCD. We study them by applying holographic techniques. Chapter 8 details the construction of a bottom-up type holographic theory for QCD that involves magnetic phenomena and describes its applications to open problems in this field of research.

One general problem concerns the phase diagram of QCD in the presence of external magnetic fields, see Fig. 1.1. Some important questions regarding the phase diagram involve: how do the phase boundaries at $B = 0$ extend to non-vanishing B? Are there new phases triggered by a non-trivial B? how do the thermodynamic potentials and the transport properties of the system are affected by magnetic fields? Are there new transport mechanisms induced by the presence of magnetic fields? These questions are relevant both for heavy-ion collisions where intense magnetic fields are generated (see Fig. 1.3) in the off-central collisions and astrophysical objects as neutron stars and magnetars that are known to have large magnetic fields.

Another important question concerns dependence of the QCD vacuum state on magnetic fields. The vacuum is characterized by a non-trivial quark condensate. Using field theory techniques, it can be shown that, for small temperatures a small applied magnetic field enhances this condensate. This is called *magnetic catalysis* [15–17], and, is confirmed by lattice simulations [18]. However, the same lattice simulations also observe the inverse behavior, i.e. decrease in the magnitude of the condensate for larger values of the temperature and magnetic fields [19, 20]. This is called *inverse magnetic catalysis* and its physical origin is still an open question. As we show in Chap. 8 holography provides important insights into this problem.

Finally, a novel class of charge, energy and chirality transport mechanism is induced by the magnetic fields. This is called *anomalous transport* with the prime example being the *chiral magnetic effect* [21–23]. In short, transitions in the vacuum state of QCD between states with topologically different gluon configurations induce a non-trivial electric current along an external magnetic field. How to characterize the magnitude of this transport—that depends on the value of *chiral magnetic conductivity*—is an open problem where holography provides an alternative approach [24, 25]. This and similar effects i.e. *chiral magnetic wave, chiral vortical effect* are believed to be realized in heavy ion collisions and might play an important role in solving long-standing problems as the baryon asymmetry in our universe [26].

The outline of the rest of the book is as follows. After a brief introduction to QCD, heavy ion collisions and the gauge-gravity duality in the next three chapters, we will detail the construction of the bottom-up holographic model for large N gauge theories coupled to electromagnetism. We focus on the hadron spectra, properties of the ground state, thermodynamics and transport with a special focus on the electromagnetic probes and their realization in the holographic description. We end with an outlook.

This book is not meant to be exhaustive. Among others, dynamical phenomena, for example. the formation process of the quark-gluon plasma in heavy ion collisions and black hole formation is left out. There is extensive material in the literature on the theory of heavy ion collisions, see e.g. [27] for a recent review and the references therein, magnetic phenomena in quantum field theories, see e.g. [28, 29], and applications of holography to heavy ion collisions, see e.g. [30] where one can find treatment of subjects falling out of the scope of this book.

References

1. P. de Forcrand, O. Philipsen, Nucl. Phys. B **642**, 290-306 (2002). https://doi.org/10.1016/S0550-3213(02)00626-0. arXiv:hep-lat/0205016 [hep-lat]
2. G. Aarts, E. Seiler, I.O. Stamatescu, Phys. Rev. D **81**, (2010). https://doi.org/10.1103/PhysRevD.81.054508. arXiv:0912.3360 [hep-lat]
3. L.D. Landau, E.M. Lifshitz, *Fluid Mechanics* (Pergamon, 1987)
4. R. Kubo, J. Phys. Soc. Jap. **12**, 570–586 (1957). https://doi.org/10.1143/JPSJ.12.570
5. P.F. Kolb, U.W. Heinz, arXiv:nucl-th/0305084 [nucl-th]
6. P. Romatschke, U. Romatschke, Phys. Rev. Lett. **99**, (2007). https://doi.org/10.1103/PhysRevLett.99.172301. arXiv:0706.1522 [nucl-th]
7. C. Shen, U. Heinz, P. Huovinen, H. Song, Phys. Rev. C **84** (2011). https://doi.org/10.1103/PhysRevC.84.044903. arXiv:1105.3226 [nucl-th]
8. C. Gale, S. Jeon, B. Schenke, Int. J. Mod. Phys. A **28**, 1340011 (2013). https://doi.org/10.1142/S0217751X13400113. arXiv:1301.5893 [nucl-th]
9. C. Gale, S. Jeon, B. Schenke, P. Tribedy, R. Venugopalan, Phys. Rev. Lett. **110** (2013) no.1, 012302 https://doi.org/10.1103/PhysRevLett.110.012302. arXiv:1209.6330 [nucl-th]
10. P. Romatschke, Int. J. Mod. Phys. E **19**, 1–53 (2010). https://doi.org/10.1142/S0218301310014613. arXiv:0902.3663 [hep-ph]
11. J.M. Maldacena, Int. J. Theor. Phys. **38**, 1113 (1999) [Adv. Theor. Math. Phys. **2** (1998) 231] https://doi.org/10.1023/A:1026654312961 [hep-th/9711200]
12. S.S. Gubser, I.R. Klebanov, A.M. Polyakov, Phys. Lett. B **428**, 105–114 (1998). https://doi.org/10.1016/S0370-2693(98)00377-3. arXiv:hep-th/9802109 [hep-th]
13. E. Witten, Adv. Theor. Math. Phys. **2**, 253–291 (1998). https://doi.org/10.4310/ATMP.1998.v2.n2.a2. arXiv:hep-th/9802150 [hep-th]

14. D.E. Kharzeev, Prog. Part. Nucl. Phys. **75**, 133–151 (2014). https://doi.org/10.1016/j.ppnp.2014.01.002. arXiv:1312.3348 [hep-ph]

15. V.P. Gusynin, V.A. Miransky, I.A. Shovkovy, Phys. Rev. Lett. **73**, 3499 (1994) Erratum: [Phys. Rev. Lett. **76** (1996) 1005] https://doi.org/10.1103/PhysRevLett.73.3499 [hep-ph/9405262]

16. V.P. Gusynin, V.A. Miransky, I.A. Shovkovy, Phys. Lett. B **349**, 477 (1995). https://doi.org/10.1016/0370-2693(95)00232-A [hep-ph/9412257]

17. V.P. Gusynin, V.A. Miransky, I.A. Shovkovy, Phys. Rev. D **52**, 4718 (1995). https://doi.org/10.1103/PhysRevD.52.4718 [hep-th/9407168]

18. G.S. Bali, F. Bruckmann, G. Endrodi, Z. Fodor, S.D. Katz, S. Krieg, A. Schafer, K.K. Szabo, JHEP **1202**, 044 (2012). https://doi.org/10.1007/JHEP02(2012)044. arXiv:1111.4956 [hep-lat]

19. G.S. Bali, F. Bruckmann, G. Endrodi, Z. Fodor, S.D. Katz, S. Krieg, K.K. Szabo, PoS LATTICE **2011**, 192 (2011). arXiv:1111.5155 [hep-lat]

20. G.S. Bali, F. Bruckmann, G. Endrodi, Z. Fodor, S.D. Katz, A. Schafer, Phys. Rev. D **86**, (2012). https://doi.org/10.1103/PhysRevD.86.071502. arXiv:1206.4205 [hep-lat]

21. D.E. Kharzeev, L.D. McLerran, H.J. Warringa, Nucl. Phys. A **803**, 227 (2008). https://doi.org/10.1016/j.nuclphysa.2008.02.298. arXiv:0711.0950 [hep-ph]

22. K. Fukushima, D.E. Kharzeev, H.J. Warringa, Phys. Rev. D **78**, (2008). https://doi.org/10.1103/PhysRevD.78.074033. arXiv:0808.3382 [hep-ph]

23. D.E. Kharzeev, J. Liao, S.A. Voloshin, G. Wang, Prog. Part. Nucl. Phys. **88**, 1–28 (2016). https://doi.org/10.1016/j.ppnp.2016.01.001. arXiv:1511.04050 [hep-ph]

24. J. Erdmenger, M. Haack, M. Kaminski, A. Yarom, JHEP **01**, 055 (2009). https://doi.org/10.1088/1126-6708/2009/01/055. arXiv:0809.2488 [hep-th]

25. K. Landsteiner, E. Megias, F. Pena-Benitez, Lect. Notes Phys. **871**, 433–468 (2013). https://doi.org/10.1007/978-3-642-37305-3_17. arXiv:1207.5808 [hep-th]

26. D.E. Kharzeev, J. Liao, Nature Rev. Phys. **3**(1), 55–63 (2021). https://doi.org/10.1038/s42254-020-00254-6. arXiv:2102.06623 [hep-ph]

27. W. Busza, K. Rajagopal, W. van der Schee, Ann. Rev. Nucl. Part. Sci. **68**, 339–376 (2018). https://doi.org/10.1146/annurev-nucl-101917-020852. arXiv:1802.04801 [hep-ph]

28. D.E. Kharzeev, K. Landsteiner, A. Schmitt, H.U. Yee, Lect. Notes Phys. **871**, 1–11 (2013). https://doi.org/10.1007/978-3-642-37305-3_1. arXiv:1211.6245 [hep-ph]

29. V.A. Miransky, I.A. Shovkovy, Phys. Rept. **576**, 1 (2015). https://doi.org/10.1016/j.physrep.2015.02.003. arXiv:1503.00732 [hep-ph]

30. J. Casalderrey-Solana, H. Liu, D. Mateos, K. Rajagopal, U.A. Wiedemann, https://doi.org/10.1017/CBO9781139136747. arXiv:1101.0618 [hep-th]

Chapter 2
Basic Features of QCD

2.1 Lagrangian

We start with a reminder of basic definitions and salient features of QCD.[1] Strong
nuclear force is arguably the least understood among the four fundamental forces
in the universe. This is the force that binds the protons and neutrons in the atomic
nuclei, holding them together. This also hints at the reasons underlying our incomplete
understanding of the strong force: it is confined to subatomic distances, length scales
of order 1 fm and time scales of order 1 fm/c.[2]

Strong force carries a specific type of charge, called *color charge*. Just as electric
charge is associated to the U(1) phase rotations of quantum fields, color is associated
to non-Abelian phase rotations of SU(3) gauge group. Fundamental representations
of this gauge group, *quarks*, are fermions. Along with color, quarks also carry electric
charge. Denoting the SU(3) generators by $T^a, a = 1, \ldots 8$, transformations of quarks
under color and electric rotations are respectively given by

$$q_i \to e^{i\alpha^a T^a} q_i \qquad q_i \to e^{ie\eta_i} q_i,$$
(2.1)

where i is the flavor index running over the following 6 type of quarks: u ($\sim$5 MeV),
c ($\sim$1.5 GeV), t ($\sim$175 GeV), d ($\sim$10 MeV), s ($\sim$100 MeV), b ($\sim$5 GeV) where we
also showed their respective masses, and the constant $\eta_i = +2/3$ for $i = 1, 2, 3$ and
$-1/3$ for $i = 4, 5, 6$. The remaining canonical fields are the 8 gluons that are in the
adjoint representation of the gauge group. The lagrangian of QCD is given by

$$\mathcal{L}_{QCD} = -\frac{1}{4} F^a_{\mu\nu} F^{a,\mu\nu} + \theta \epsilon^{\mu\nu\alpha\beta} F^a_{\mu\nu} F^a_{\alpha\beta} + i \sum_{i=1}^{6} \bar{q}_i \gamma^\mu (\partial_\mu - ig A^a_\mu T^a + m_i) q_i,$$
(2.2)

---

[1] I assume that the reader is somewhat familiar with QCD.

[2] c denotes the speed of light which we set as $c = 1$ throughout the book.

© The Author(s), under exclusive license to Springer Nature Switzerland AG 2021
U. Gürsoy, *Holography and Magnetically Induced Phenomena in QCD*,
SpringerBriefs in Physics,
https://doi.org/10.1007/978-3-030-79599-3_2

9

where A_μ^a are the gluons and their field strength F contains gluon self interactions

$$F_{\mu\nu}^a = \partial_\mu A_\nu^a - \partial_\nu A_\mu^a + g f^{abc} A_\mu^b A_\nu^c, \tag{2.3}$$

with g the QCD coupling constant, fully antisymmetric f^{abc} are the SU(3) structure constants, and $\theta \sim \theta + 2\pi$ is the QCD theta-angle. To complete the definition of the quantum theory one needs two additional sectors which we will not write explicitly: (1) to define the path integral over gluons one introduces a Faddeev–Popov determinant that can be represented by the additional ghost fields, (2) to remove divergences from loops one adds a counterterm action whose coefficients are fixed by renormalization conditions. We will not go into details of renormalization here, but only mention an important consequence: As with any renormalizable QFT, renormalization procedure renders the coupling constant g dependent on the RG scale. We discuss this below. The constant θ, on the other hand, is topological and (perturbatively) not renormalized.

2.1.1 Asymptotic Freedom

The dependence of QCD coupling constant g on the energy scale is determined by the beta-function equation

$$\frac{dg}{d \log Q} = \beta(g). \tag{2.4}$$

A first distinctive feature of QCD is that $\beta < 0$. This means that the theory becomes strongly interacting in the IR, and becomes free in the UV. Discovery of this property, called *asymptotic freedom*, led to the 2004 Nobel prize in physics (Gross, Politzer, Wilczek) [1, 2]. This behavior is the opposite of QED, as well as many other QFTs which become free in the IR. For a detailed and recent study of the QCD coupling constant, including the non-perturbative regime, see [3].

2.2 Confinement

Another fundamental property of QCD is the fact that quarks and gluons, namely particles that carry color charge, are only found in colorless combinations i.e. *baryons* such as protons and neutrons, *mesons* such as pions and kaons and *glueballs*.[3]

This is called *confinement*, a property whose theoretical origin is still unclear in the sense that its proof based on the microscopic theory (2.2) is yet to be established.

---

[3] These are colorless combinations comprised of only gluons such as the excitations created by the operator $F_{\mu\nu}^a F^{a,\mu\nu}$. They are observed in lattice simulations, see e.g. [4, 5], but hard to detect due to short lifetime.

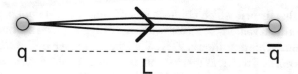

Fig. 2.1 Linear confinement of a quark-antiquark pair arising from a gluon flux tube. The linear behavior of the potential can be understood by applying Gauss' law assuming that the gluon flux lines are confined inside a narrow string

We will see in Chap. 5 that holography provides a useful alternative perspective on confinement.

Confinement of a quark and an antiquark within a meson state follows from an interaction potential that is linearly growing with distance. This quark-antiquark potential can be expressed as

$$V_{q\bar{q}}(L) \approx \sigma_0 L - \frac{\alpha_{eff}}{L}, \tag{2.5}$$

where σ_0 is called the string tension and α_{eff} is an effective QCD coupling. The second term in (5.23) is the analog of Coulomb interaction between particles of two opposite charges, while the first term can be understood as arising from a gluon flux tube stretching like a string between the quark and the anti-quark (see Fig. 2.1). This is termed *linear confinement*.

The theory in the absence of quarks, that is pure SU(N) Yang–Mills theory, also confines color charge. In this case the colorless asymptotic states are glueballs. In fact, confinement can be characterized more clearly in the absence of fundamental matter. Consider Yang–Mills theory at finite temperature. Whether this theory confines color charge or not can be determined by adding a test quark and measuring its free energy. Just like electric charge on a sphere would have infinite energy (because the electric field it emanates cannot terminate anywhere), a test quark in a confining state would also have infinite energy. This is measured by the vacuum expectation value (VeV) of the *Polyakov loop* operator

$$\mathcal{P} = \frac{1}{N} \langle \text{Tr} \, P \, e^{-\oint A_0^a T^a} \rangle, \tag{2.6}$$

where P stands for path ordering and the integral is taken over the Euclidean time circle. The exponent is the action of a static test quark and it diverges, as argued, in the confined state. Hence $\mathcal{P} = 0$ when the theory confines. Notice that $\mathcal{P}$ is gauge invariant, hence, it serves as a good order parameter for confinement. However, it is not invariant under the *center* element of the group. For SU(N) this center is the discrete subgroup Z_N. A non-trivial expectation value of the Polyakov loop $\mathcal{P} = 0$ would signal spontaneous breaking of this center symmetry. Therefore, in pure Yang–Mills (as well as Yang–Mills with adjoint matter) spontaneous breaking of center of the gauge group is associated to the deconfinement phase transition. It turns out that this theory, which confines at sufficiently low temperatures ceases to do so above a critical value $T > T_c$ for which the Polyakov loop acquires a nontrivial expectation value. This transition at T_c is called the *confinement-deconfinement phase transition*. It is not at all obvious that a theory that confines at low temperatures should deconfine at higher temperatures [6]. Later in Chap. 5 we will see that holographic correspondence provides a simple and universal explanation of this fact.

2.3   Chiral Symmetry and Axial Anomaly

Another salient feature of QCD, which will play an important role in the next chapters is the chiral symmetry of the lagrangian (2.2) in the limit of vanishing quark masses, $m_i = 0$. In this limit, chirality is classically conserved as the quark kinetic terms in (2.2) separate into left and right-handed components $q_{L,R} \equiv \frac{1}{2}(1 \pm \gamma_5)q$. As a result, the classical theory acquires the following global symmetry

$$q_{L,i} \to (U_L)_i^j \, q_{L,j}, \qquad\qquad q_{R,i} \to (U_R)_i^j \, q_{R,j} \,, \qquad\qquad (2.7)$$

where $U_{L,R}$ are separate unitary matrices. For generality, let us assume N_f different quark flavors. Thus, in the massless limit the continuous global symmetry of the theory becomes $U(N_f) \times U(N_f)$. This symmetry pattern changes in the quantum theory for two separate reasons. First, there is an axial anomaly in the $U(1)$ subgroup U_{L-R} where the subscript refers to the linear combination of the generators of the left and right $U(N_f)$ groups. This anomaly is expressed as non-conservation of the axial current $J_\mu^5 \equiv \sum_i \bar{q}_i \gamma^5 \gamma_\mu q_i$ [7, 8]

$$\partial_\mu J_\mu^5 = -\frac{g^2 N_f}{32\pi^2} \epsilon^{\mu\nu\alpha\beta} F_{\mu\nu}^a F_{\alpha\beta}^a \,. \qquad\qquad (2.8)$$

The remaining global symmetry of the theory $SU(N_f)_{L+R} \times SU(N_f)_{L-R} \times U(1)_{L+R}$ is not anomalous.

   The second reason the original symmetry of the classical lagrangian changes is that, part of it is spontaneously broken due to presence of a non-trivial quark condensate $\langle \bar{q}q \rangle = \langle \bar{q}_L q_R + \bar{q}_R q_L \rangle \neq 0$ in the quantum theory. This breaks the symmetry down to the subgroup $U_L = U_R$. The corresponding $N_f^2 - 1$ broken generators of $SU(N_f)_{L-R}$ give rise to would-be-massless Goldstone bosons, which in reality, correspond to the low-mass mesons in the theory, i.e. pions, kaons etc. The reason that these in fact are massive in real QCD is that quarks have mass to start with, therefore the Goldstone bosons are not massless but light, and called *pseudo-Goldstone bosons*. In fact, in QCD only the masses of u and the d quarks are small, hence one typically only considers three lowest mass pseudo-Goldstone bosons, i.e. the pions π^0, π^+ and π^-.

   In passing, we note that this symmetry pattern changes altogether when QCD is coupled to electromagnetism because of the different charges of the up-type (u, c, t) and the down-type (d, s, b) quarks. In the general case of $N_f = N_u + N_d$ flavors with N_u up-type and N_d down-type quarks, the remaining symmetry is $SU(N_u) \times SU(N_d) \times U(1)$.

2.4 QCD Thermodynamics

QCD at finite temperature has a relatively rich structure, see [6] for a review. For vanishing quark chemical potential, the thermodynamic functions can be directly determined from lattice simulations[4] by imposing periodic boundary conditions on Euclidean time direction with perimeter given by inverse temperature $1/T$. This is an extensive system, with pressure given by the free energy density

$$p = -f = Ts - \varepsilon, \tag{2.9}$$

with p, f, s and ε denoting pressure, free energy, entropy and energy that are only functions of T (at vanishing density) in the canonical ensemble,

$$dp = sdT. \tag{2.10}$$

Due to confinement, thermodynamic functions e.g. entropy at low temperatures are dominated by hadronic degrees of freedom. In contrast, quarks and gluons dominate high temperatures. Lattice studies reveal that this change in behavior of the free energy does not lead to a phase transition in the presence of flavor (as opposed to pure Yang–Mills which exhibits a first order phase transition) but a sharp cross-over around $T_c \approx 150$ MeV, see Fig. 2.2.

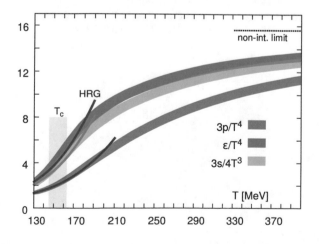

Fig. 2.2 Equation of state of QCD with $N_f = 3$ flavors [9]. "HRG" refers to hadron resonance gas model. Confinement-deconfinement crossover takes place around $T \approx 150$ MeV. Reprinted with permission from APS publishers

---

[4] We study finite chemical potential using gauge-gravity duality in Chap. 7.

At very large temperatures, the QCD equation of state approaches to the Stefan–Boltzman gas, that is a free gas of quarks and gluons, due to asymptotic freedom. In this limit the thermodyamic functions are given by [10]

$$p = \frac{1}{3}\varepsilon = \frac{1}{4}s = \left(2(N^2 - 1) + 4NN_f\frac{7}{8}\right)\frac{\pi^2 T^4}{90}. \tag{2.11}$$

The additional relations among s, ε and p arise from conformality of the theory in the high temperature limit—which asserts vanishing of the trace anomaly $\langle T^\mu_\mu \rangle = \epsilon - 3p$—and Eq. (2.10).

References

1. D.J. Gross, F. Wilczek, Phys. Rev. Lett. **30**, 1343–1346 (1973). https://doi.org/10.1103/PhysRevLett.30.1343
2. H.D. Politzer, Phys. Rev. Lett. **30**, 1346–1349 (1973). https://doi.org/10.1103/PhysRevLett.30.1346
3. A. Deur, S.J. Brodsky, G.F. de Teramond, Nucl. Phys. **90**, 1 (2016). https://doi.org/10.1016/j.ppnp.2016.04.003. arXiv:1604.08082 [hep-ph]
4. C.J. Morningstar, M.J. Peardon, Phys. Rev. D **60** (1999). https://doi.org/10.1103/PhysRevD.60.034509. arXiv:hep-lat/9901004 [hep-lat]
5. Y. Chen, A. Alexandru, S.J. Dong, T. Draper, I. Horvath, F.X. Lee, K.F. Liu, N. Mathur, C. Morningstar, M. Peardon et al., Phys. Rev. D **73** (2006). https://doi.org/10.1103/PhysRevD.73.014516. arXiv:hep-lat/0510074 [hep-lat]
6. D.J. Gross, R.D. Pisarski, L.G. Yaffe, Rev. Mod. Phys. **53**, 43 (1981). https://doi.org/10.1103/RevModPhys.53.43
7. S.L. Adler, Phys. Rev. **177**, 2426–2438 (1969). https://doi.org/10.1103/PhysRev.177.2426
8. J.S. Bell, R. Jackiw, Nuovo Cim. A **60**, 47–61 (1969). https://doi.org/10.1007/BF02823296
9. A. Bazavov et al., HotQCD. Phys. Rev. D **90**, (2014). https://doi.org/10.1103/PhysRevD.90.094503. arXiv:1407.6387 [hep-lat]
10. O. Philipsen, "The QCD equation of state from the lattice", Prog. Part. Nucl. Phys. **70**, 55–107 (2013). https://doi.org/10.1016/j.ppnp.2012.09.003

Chapter 3
Hydrodynamics and Transport Coefficients

QCD at intermediate temperatures, $130 \text{ MeV} \lesssim T \lesssim 500 \text{ MeV}$, that are relevant for heavy ion collisions, and at small densities, is well approximated by a fluid of quarks and gluons. This fluid is theoretically described by relativistic hydrodynamics. To get an idea how well hydrodynamic simulations agree with observables in heavy ion collisions, see Fig. 3.1. Hydrodynamics is a theory organized in a derivative expansion, that is an expansion in powers of frequency or momentum compared to an intrinsic scale, such as the mean free path in systems with quasi-particle excitations, $k\ell_{mfp}$, or compared to temperature k/T in systems, such as strongly correlated plasmas, where quasi-particle like description ceases to exist. Hydrodynamical variables are determined by conservation laws, such as the energy-momentum and charge conservation in the plasma, and these equations are also subjected to the derivative expansion.

3.1 Generalities

In this section, we consider hydrodynamics of neutral relativistic fluid, hence the only non-trivial conservation equation is the energy-momentum conservation:

$$\nabla_\mu T^{\mu\nu} = 0. \tag{3.1}$$

These are 4 equations for a matrix with 10 components hence the system is under-determined. To reduce the number of unknowns to 4 one resorts to the Lorentz symmetry and derive constitutive relations accordingly. For the neutral plasma the 4 fluid variables (with fixed flat metric $g_{\mu\nu} = \eta_{\mu\nu}$) can be taken as the 4-velocity u^μ and temperature T:

$$u^\mu(x), \quad g_{\mu\nu}u^\mu u^\nu = -1; \quad T(x). \tag{3.2}$$

Fig. 3.1 Very good fit of hydrodynamic simulations (solid symbols) to experimental data (open symbols) from RHIC depicting an energy flow parameter $v_2 4$ of the plasma. Figure adapted from [1]

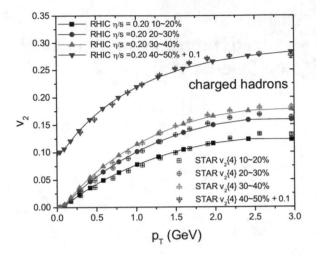

The constitutive relation that expresses $T^{\mu\nu}$ in terms of these unknowns is obtained as follows. In relativistic hydrodynamics, at zeroth order in the derivative expansion, the only symmetric two-index objects one can use to construct the stress tensor are $\eta_{\mu\nu}$ and $u^\mu u^\nu$, therefore one postulates

$$T_0^{\mu\nu} = u^\mu u^\nu (\varepsilon + p) + g^{\mu\nu} p , \tag{3.3}$$

where we parametrized the coefficients in terms of energy ε and pressure p of the fluid.[1] This term at zeroth order in the derivative expansion corresponds to *ideal* relativistic fluid. Energy and pressure as a function of temperature should be defined using the microscopic definition of the theory, such as lattice-QCD, as explained in the previous section.

The next term in the derivative expansion involves dissipative terms. For a neutral plasma[2] we have only two such terms corresponding to *shear* and *bulk* viscosities. The derivation can be found in standard textbooks and review papers[3] and this contribution is given by

$$T_1^{\mu\nu} = P^{\mu\alpha} P^{\nu\beta} \left[\eta \left(\partial_\alpha u_\beta + \partial_\beta u_\alpha - \frac{2}{3} g_{\alpha\beta} \, \partial \cdot u \right) + \zeta \, g_{\alpha\beta} \, \partial \cdot u \right] , \tag{3.4}$$

where $P^{\alpha\beta}$ projects to the plane transverse to u^α:

$$P^{\alpha\beta} = g^{\alpha\beta} + u^\alpha u^\beta , \tag{3.5}$$

---

[1] It is often useful to express quantities in the rest frame $u^\mu = (1, 0, 0, 0)$ where $T^{00} = \varepsilon$ and $T^{ii} = p$.

[2] In case of a charged relativistic fluid there is also a contribution from conductivity at this level.

[3] See e.g. [2].

and the coefficients η and ζ are called "shear viscosity" and "bulk viscosity" respectively. They characterize response of the fluid to shear (traceless) and volume (trace) deformations. One encounters more transport coefficients at higher orders in the derivative expansion. Cutting off this expansion at first order results in acausality which can be circumvented [3–5] by treating the first order energy momentum tensor as an independent dynamical valuable whose evolution is determined by the second order coefficients. This provides a very good approximation.

One should emphasize that, apart from its applications to heavy ion collisions, astrophysics etc, development of systematics of relativistic hydrodynamics itself is an extremely active research topic today. Two recent developments that are worth mentioning here are the hydrodynamic attractors [6], which indicate that hydrodynamics also apply far from equilibrium and an alternative first order formulation [7] which preserves causality.

3.2 Transport Coefficients

The transport coefficients, in our case only the shear and bulk viscosity, follow from the microscopic properties of the fluid. According to linear response theory, the first order change in the expectation value of an operator O_B due to deformation of the Lagrangian by an operator O_A is given by the retarded Green's function of O_B and O_A:

$$\mathcal{L} \to \mathcal{L} + \int O^A \delta\phi_A \quad \Rightarrow \quad \langle O^B \rangle = G_R^{BA} \delta\phi_A \,, \tag{3.6}$$

where the retarded Green's function is given by

$$G_R^{BA}(\omega, \mathbf{k}) = -i \int d^4x\, e^{-ik \cdot x} \theta(t) \langle [O^A(t, \mathbf{x}), O^B(0, \mathbf{0})] \rangle \,. \tag{3.7}$$

The final average is a thermal average. Shear and bulk viscosities characterize how the energy-momentum of a system responds to deformations in the fluid element that can be represented as metric deformations. As the metric deformations themselves couple to the energy-momentum tensor, both O_A and O_B above are given by components of $T_{\mu\nu}$ and the shear and the bulk viscosities are obtained in the limit

$$\eta \left(\delta^{il} \delta^{km} + \delta^{im} \delta^{kl} - \frac{2}{3} \delta^{ik} \delta^{lm} \right) + \zeta \delta^{ik} \delta^{lm} = \lim_{\omega \to 0} \frac{i}{\omega} G_R^{ik,lm}(\omega, \mathbf{0}) \,, \tag{3.8}$$

with momentum $\mathbf{k}$ set to zero. Thus, the shear viscosity can be read off e.g. from the $(12, 12)$ and the bulk viscosity can be read off from the $(11 + 22 + 33, 11 + 22 + 33)$ components of the Green's function of the energy-momentum tensor.

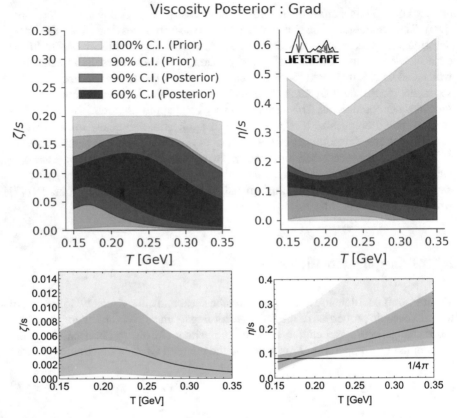

Fig. 3.2 Shear and bulk viscosity of QCD normalized by entropy as a function of temperature obtained by two independent Bayesian analyses of data from heavy ion collisions. Top figures from [13], bottom figures from [14, 15]

These correlators can be computed using perturbation theory at weak coupling. For the shear viscosity of QCD in the large-N limit one finds [8]

$$\frac{\eta}{s} \propto -\frac{1}{\lambda^2 \log \lambda},$$ (3.9)

where λ is the 't Hooft coupling and s is the entropy density. There is a similar formula for the bulk viscosity [9]. However, as we discussed earlier the weak coupling approximation becomes invalid at low temperatures in the quark-gluon plasma created in heavy ion collisions and one has to resort to non-perturbative techniques.

Computation of transport coefficients in QCD is an open problem and there exist at least four other approaches. (i) One is kinetic theory [10] which treats the quark-gluon system at high temperatures as a weakly interacting gas of particles. This does not, however necessarily apply at intermediate temperatures when its fluid-like properties become evident. (ii) Another approach is lattice QCD. We explained above that lattice is not suitable for real-time dissipative processes but one may still compute the Euclidean analogs of the correlation functions (3.7) reliably on the lattice, and try to analytically continue them to real time. However, this analytic continuation procedure requires knowledge of the entire density of states of QCD, for which one can at best make an ansatz [11, 12]. This typically results in large systematic errors. (iii) The third method is Bayesian analysis of data generated in heavy ion collisions. Here, one reverse engineers the data using hydrodynamic simulations and obtains an optimal fit of transport coefficients entering as input in these simulations, to as many experimental observables as possible. This is perhaps the most systematic of all approaches and it is believed to provide at least the right ballpark, see Fig. 3.2 for the latest updates [13–15]. However, it is still not free of systematic uncertainties which stem from the possibility of different set of variables one can use in optimization, initial conditions for the simulations and characterization of the hadronization procedure [14, 15]. Also, as this is not a first-principles calculation, it somewhat obscures the theoretical understanding of the physics involved. (iv) Finally, gauge-gravity duality provides the correct answer for the incorrect theory i.e. maximally supersymmetric Yang-Mills theory at strong coupling and large N. However, certain arguments of universality of transport at strong coupling can be used to derive valuable results for QCD at large N as well. There are also reasons to believe that the corrections due to large N approximation are small, see Sect. 5. The latter is the method we focus on in this book and present the holographic computation of shear viscosity in the next chapter. Perhaps an overarching approach encompassing and extrapolating between all these techniques would be the best way forward.

References

1. U. Gürsoy, D. Kharzeev, E. Marcus, K. Rajagopal and C. Shen, Phys. Rev. C **98**(5), 055201 (2018). https://doi.org/10.1103/PhysRevC.98.055201. arXiv:1806.05288 [hep-ph]
2. P. Romatschke, Int. J. Mod. Phys. E **19**, 1 (2010). https://doi.org/10.1142/S0218301310014613. arXiv:0902.3663 [hep-ph]
3. I. Muller, Z. Phys. **198**, 329–344 (1967). https://doi.org/10.1007/BF01326412
4. W. Israel, Ann. Phys. **100**, 310–331 (1976). https://doi.org/10.1016/0003-4916(76)90064-6
5. W. Israel, J.M. Stewart, Ann. Phys. **118**, 341–372 (1979). https://doi.org/10.1016/0003-4916(79)90130-1
6. M.P. Heller, M. Spalinski, Phys. Rev. Lett. **115**(7), 072501 (2015). https://doi.org/10.1103/PhysRevLett.115.072501. arXiv:1503.07514 [hep-th]
7. P. Kovtun, JHEP **10**, 034 (2019). https://doi.org/10.1007/JHEP10(2019)034. arXiv:1907.08191 [hep-th]
8. A. Hosoya, K. Kajantie, Nucl. Phys. B **250**, 666–688 (1985). https://doi.org/10.1016/0550-3213(85)90499-7
9. P.B. Arnold, C. Dogan, G.D. Moore, Phys. Rev. D **74**, (2006). https://doi.org/10.1103/PhysRevD.74.085021. arXiv:hep-ph/0608012 [hep-ph]
10. P.B. Arnold, G.D. Moore, L.G. Yaffe, JHEP **01**, 030 (2003). https://doi.org/10.1088/1126-6708/2003/01/030. arXiv:hep-ph/0209353 [hep-ph]

11. H.B. Meyer, Phys. Rev. D **76**, (2007). https://doi.org/10.1103/PhysRevD.76.101701. arXiv:0704.1801 [hep-lat]
12. H.B. Meyer, Phys. Rev. Lett. **100**, (2008). https://doi.org/10.1103/PhysRevLett.100.162001. arXiv:0710.3717 [hep-lat]
13. D. Everett et al. [JETSCAPE]. arXiv:2011.01430 [hep-ph]
14. G. Nijs, W. Van Der Schee, U. Gürsoy, R. Snellings, arXiv:2010.15134 [nucl-th]
15. G. Nijs, W. van der Schee, U. Gürsoy, R. Snellings, arXiv:2010.15130 [nucl-th]

Chapter 4
Gauge-Gravity Duality

Gauge-gravity duality provides an alternative tool to study strongly interacting gauge theories by mapping quantum systems to semi-classical gravity in higher dimensions. In this book, we apply the gravitational description as an approximation to compute n-point functions of large-N QCD in the limit of infinite coupling.

Essence of the gauge-gravity correspondence is an equivalence between open and closed string descriptions of scattering processes in backgrounds with D-branes [1, 2] in string theory. In the particular case of absorption/emission of string states from D3 branes in IIB string theory this open/closed duality was made precise by Maldacena [3] in 1997. I will reserve "AdS/CFT correspondence" for this specific example, which was later generalized to theories that are more akin to gauge theories describing particle physics. Prescription to compute n-point functions in the QFT was first clarified in the works [4, 5] for the maximally supersymmetric Yang-Mills conformal field theory in 4D. However, the basic prescription remains the same for generic QFTs, as we explain below.

It is important to note however, that, the duality is quite likely a more fundamental property of quantum gravity [6–9]—and separately a fundamental property of certain quantum field theories thereof[1]—which goes beyond string theory. This more generic point of view provides the main motivation for the bottom-up holographic models for QCD in the next chapter.

In passing let us note that the gauge-gravity duality, when applied in the opposite direction, also serves as a definition of quantum gravity in asymptotically AdS space-times. This, recently lead to advanced understanding of structure of spacetime emerging from entanglement between quantum states [11, 12] and the black hole information paradox [13, 14] to mention a few.

---

[1] See for example [10] for how free field n-point functions can be reformulated as gravity amplitudes.

© The Author(s), under exclusive license to Springer Nature Switzerland AG 2021
U. Gürsoy, *Holography and Magnetically Induced Phenomena in QCD*,
SpringerBriefs in Physics,
https://doi.org/10.1007/978-3-030-79599-3_4

4.1   Holographic Computation of Correlators

Consider generating function of a QFT formally obtained by integrating out the canonical fields in the presence of a source $S(x)$ for a gauge-invariant operator $\mathcal{O}(x)$

$$e^{i\mathcal{W}[J(x)]} = \int \mathcal{D}\phi \, e^{i \int \mathcal{L}[\phi] + S(x)\mathcal{O}(x)} . \tag{4.1}$$

Connected correlators of $\mathcal{O}$ are computed from $\mathcal{W}[S(x)]$ by varying with respect to J. Just like any coupling constant, one can think of S acquiring a renormalization energy scale dependence under renormalization group, becoming $S(x, Q)$ with $S(x)$ in (4.1) being its value at the UV cut-off. If this QFT emerges from the low energy excitations of a collection of D-branes, then (4.1) corresponds to the open string description. In the dual closed string description, the energy scale Q is mapped onto a true "holographic" coordinate r. In this description, the source extends into the bulk, whereby $S(x, Q)$ is represented as a dynamical closed string state $\phi(x, r)$ with $S(x)$ in (4.1) being its value on the boundary $r = r_0$. $\mathcal{W}[S(x)]$ in the closed string description is then obtained by solving the equation of motion of this closed string field, and, substituting it in the string field theory action on-shell:

$$\mathcal{W}[J(x)] = \log Z_{string}[\phi(x, r \to r_0) \to S(x)] . \tag{4.2}$$

The detailed map between quantum field theory and the closed string theory is best understood in the original example of [3], where the open string side is given by $\mathcal{N} = 4$ super Yang-Mills theory and the closed string side is IIB string theory on the $AdS_5 \times S^5$ background. Next well understood cases in 4D are theories obtained from $\mathcal{N} = 4$ sYM by relevant or marginal deformations. In such theories, generally, there exists the following correspondence between the parameters on the two sides:

$$g_s \sim g^2, \qquad R\ell_s^2 \sim (g^2 N)^{-\frac{1}{2}} , \tag{4.3}$$

where g_s is the string coupling constant and $R\ell_s^2$ is the Ricci curvature of the gravitational background in string length unit, g is the Yang-Mills coupling constant, and N the rank of the gauge group (number of colors). A simple limit of the AdS/CFT is obtained by the 't Hooft limit [15]:

$$N \to \infty, \quad g \to 0, \quad \lambda \equiv g^2 N \gg 1 , \tag{4.4}$$

where the combination λ is the 't Hooft coupling. This limit achieves the following, at once: (1) it removes the complications arising from string interactions by sending g_s to zero. As Newton's constant $G_N \propto g_s^2 \to 0$, the string partition function on the RHS of (4.2), in this limit, is given by the contribution of the dominant saddle. In other words, the path integral over string fields is replaced by the effective string theory (supergravity + higher derivative corrections) action on $AdS_5 \times S^5$; (2) it reduces the string theory—with arbitrary number of higher derivative terms in its effective action—to two-derivative Einstein's gravity coupled to matter fields by making curvature small. Therefore, the RHS of (4.2) further reduces to the on-shell gravity action; (3) It focuses on the strong coupling limit of gauge theory that is the non-perturbative regime where the ordinary perturbative field theory methods fail. In the rest of this book, we will only consider this particular corner of the parameter space.

In this limit, then, we have a much simpler prescription

$$\lim_{\lambda \to \infty} \lim_{N \to \infty} W[J(x)] = S_{gravity}[\phi(x, r \to r_0) \to S(x)], \qquad (4.5)$$

where $S_{gravity}$ is the gravity action on $AdS_5 \times S^5$.

An important aspect of the correspondence is that the isometries of the gravity background correspond to the global symmetries of the dual QFT. In the case of an asymptotically AdS_5 space the isometry group $SO(2, 4)$ precisely matches the conformal group[2] indicating that the dual QFT is conformally invariant. For *asymptotically AdS_5* space these isometries are realized asymptotically near the boundary indicating that the dual QFT becomes asymptotically conformal in the UV.[3] The operator $\mathcal{O}$ in (4.1) in this conformal regime is characterized by the scaling dimension Δ which is related to the mass of the scalar field[4] ϕ as

$$m^2 = \Delta(\Delta - 4). \qquad (4.6)$$

To obtain the RHS of (4.5) one solves the scalar field equation

$$\frac{1}{\sqrt{g}} \partial_\mu \sqrt{g} g^{\mu\nu} \partial_n \phi + m^2 \phi = 0 \qquad (4.7)$$

in an asymptotically AdS space-time, with the near boundary metric

$$ds^2 = \frac{1}{r^2} \left(dr^2 + \eta_{mn} dx^m dx^n \right), \qquad (4.8)$$

with η_{mn} 4D Minkowski metric. One finds two distinct asymptotic behaviors as $r \to 0$

---

[2] The isometry group of the S^5 factor, $SO(6)$ also matches the R-symmetry of the maximally supersymmetric Super Yang-Mills theory. This additional symmetry will not play any role in holographic dual of QCD in the next chapter and we will ignore it in this book.

[3] This class includes QCD which is asymptotically free with a gaussian fixed point in the UV. Holographic description of QCD all the way up to the UV fixed point would require inclusion of arbitrarily many higher derivative corrections in the gravity dual. To avoid these corrections, the theory should be cut off at a UV scale r_0 and the UV running of QCD coupling should be used as boundary conditions for $r > r_0$ which is assumed to be described by 2-derivative gravity.

[4] Similar relations for fermions, vector fields and other representations have been derived, see e.g. [16].

$$\phi(r, x) \to r^\Delta S(x) (1 + \cdots) + r^{4-\Delta} R(x) (1 + \cdots) \tag{4.9}$$

where ellipsis denote subleading terms of $O(r^2)$. We learn that the boundary condition in (4.5) should be defined more precisely as $\lim_{r \to r_0} \phi(r, x) = r_0^\Delta S(x)$. Dimensional analysis then indicates that the subleading coefficient $R(x)$ should be associated with the VeV of the operator $R(x) \propto \langle \mathcal{O}(x) \rangle$.

The full solution of the second order differential equation (4.7) requires another boundary condition which arises from the requirement of regularity in the IR $\lim_{r \to \infty} \phi(r, x) = 0$. This boundary condition corresponds to the vacuum state in the dual QFT and determines the VeV $R(x)$ in (4.9) in terms of the source $S(x)$. Using these two boundary conditions one can solve (4.7) and read off the two-point function of $\mathcal{O}$ from the ratio $R(x)/S(x)$ in the limit $S \to 0$ from (4.9).

The aforementioned IR boundary condition is modified in a thermal state. In particular, a thermal state correspond to presence of a black hole at the origin with a horizon at $r = r_h$. The AdS-Schwarzschild black hole in 5D in Poincare patch coordinates reads

$$ds^2 = \frac{1}{r^2} \left(\frac{dr^2}{1 - \frac{r^4}{r_h^4}} - dt^2 \left(1 - \frac{r^4}{r_h^4} \right) + \delta_{ij} dx^i dx^j \right), \tag{4.10}$$

with $i, j = 1, 2, 3$. In this geometry the regularity of ϕ at the origin is replaced with an infalling (outgoing) boundary condition at $r = r_h$:

$$\phi(r, x) \to (r_h - r)^{\pm i \frac{\omega}{4\pi T}}, \qquad r \to r_h, \tag{4.11}$$

where T is the Bekenstein-Hawking temperature of the black hole[5] and the $-(+)$ signs in the exponent are the infalling (outgoing) excitations at the horizon, corresponding to to boundary condition for the retarded (advanced) correlator in the thermal state of the dual field theory.

This procedure we outlined in this section, to calculate two-point correlators, more generally n-point functions, is schematically described in Fig. 1.2 where the roles of bulk field $\phi(x, r)$ and the one point function $\langle \mathcal{O}(x) \rangle$ are played by A and J in the figure. In Sect. 4.2, we work out a concrete example using this holographic prescription and compute the shear viscosity of the quark-gluon plasma.

---

[5] Throughout the book we will consider black holes with flat horizon, sometimes referred to as black branes.

4.2 An Example: Holographic Computation of Shear Viscosity

In this section, we reproduce the famous AdS/CFT result of the viscosity over entropy ratio $\frac{\eta}{s} = \frac{1}{4\pi}$ for a strongly coupled plasma [17]. As explained in Sect. 3.2, shear viscosity can be computed from the Kubo formula

$$\eta = -\lim_{\omega \to 0} \frac{1}{\omega} \mathrm{Im} G_{x_1 x_2, x_1 x_2}(\omega, 0) \tag{4.12}$$

where $G_{x_1 x_2, x_1 x_2}(\omega, 0)$ is the retarded Green's function of the $x_1 x_2$ component of the energy momentum tensor at zero momentum defined as

$$G_{\mu\nu,\alpha\beta}(\omega, k) = -i \int dt dx e^{i\omega t - ik \cdot x} \langle T_{\mu\nu}(t, x) T_{\alpha\beta}(0, 0)\rangle \tag{4.13}$$

The two point function can be computed from the holographic prescription outlined above: We assume a strongly coupled plasma at finite temperature dual to a given gravitational theory. The gravitational solution dual to this thermal state will generically be a five dimensional blackhole. To compute the two point function in (4.13) we can consider spin 2 metric fluctuations on top of this blackhole background

$$\delta g_{x_1 x_2} = \int \frac{d\omega dk}{(2\pi)^2} \phi(r) e^{-i\omega t + ikx_3}. \tag{4.14}$$

We note that the equation of motion for these fluctuations obtained by fluctuating the Einstein equations and is identical to that of a massless scalar field

$$\frac{1}{\sqrt{-g}} \partial_\mu \left[\sqrt{-g} g^{\mu\nu} \partial_\nu \delta g_{x_1 x_2}(r, t, x_3) \right] = 0. \tag{4.15}$$

For the gravitational background, we will assume a generic 5D black hole metric

$$ds^2 = e^{2A} \left[\frac{dr^2}{f} - f dt^2 + dx_i dx^i \right]. \tag{4.16}$$

We assume that this reduces to the AdS-Schwarzschild black hole (4.10) near the boundary $r \to 0$. Note that this is the most general metric consistent with the rotational symmetry on the 3D subspace spanned by x^i. Using (4.16) together with (4.14), one finds that (4.15) becomes

$$\phi'' + \left(3A' + \frac{f'}{f} \right) \phi' + \frac{\omega^2 - fk^2}{f^2} \phi = 0 \tag{4.17}$$

for the perturbation ϕ. It is easy to see that, near the boundary, Eq. (4.17) admits the following asymptotic solution

$$\phi \sim S(\omega, k) + R(\omega, k)r^4 + \cdots \tag{4.18}$$

Using the holographic prescription, one finds that the retarded Green's function is determined from the coefficients $S(\omega, k)$ and $R(\omega, k)$ by

$$G_{x_1 x_2, x_1 x_2} = \frac{1}{4\pi G} \frac{R(\omega, k)}{S(\omega, k)}, \tag{4.19}$$

where G is Newton's constant in 5D. Instead of explicitly solving (4.17) to find the coefficients S and R we consider the following simpler route [18]. Noting that viscosity in (4.12) arises from the zero frequency limit, we analyze the zero momentum and low frequency behavior of Eq. (4.17) and compare it to its near horizon behavior to read of the coefficients S and R. From now on we will set $k = 0$. Then, Eq. (4.17) has the solution

$$\phi(r) = c_1 + c_2 \int_0^r \frac{e^{-3A}}{f}, \tag{4.20}$$

in the low frequency limit. Since the constant c_2 should vanish in the exact $\omega = 0$ limit, by analyticity it should be proportional to ω for small ω. Now, expanding (4.20) near the AdS boundary, one finds that $S(\omega) = c_1$ and $R(\omega) = -\frac{c_2}{4}$.

On the other hand, near the blackhole horizon $r = r_h$ where $f(r_h) = 0$, Eq. (4.17) takes the following leading form

$$\phi'' + \frac{1}{(r - r_h)}\phi' + \frac{\omega^2}{f'(r_h)^2(r - r_h)^2}\phi = 0. \tag{4.21}$$

Then, using an ansatz of the form $\phi = (r - r_h)^\gamma (1 + ...)$ one shows that (4.21) admits the two solutions

$$\phi(r) = (r - r_h)^{\pm i\omega/f'(r_h)} F(r, \omega) \tag{4.22}$$

with $F(r, \omega, k)$ some regular function at the horizon. To determine the correct sign of (4.22) which should be of an infalling wave at the horizon, we make the following change of coordinates

$$dr^* = \frac{dr}{f} \tag{4.23}$$

$$dv_\pm = dt \pm dr^* \tag{4.24}$$

and rewrite the metric (4.16) as

$$ds^2 = e^{2A} \left(-f dv_{\pm}^2 \pm 2 dr dv_{\pm} + dx_i dx^i \right).$$ (4.25)

We note that (4.25) has a regular behavior at the horizon unlike (4.16). The corresponding light cone coordinates $v_{\pm}$ are known as outgoing and ingoing Eddington-Finkelstein coordinates. Restoring the time dependence on the perturbation and using the two solutions (4.22) together with the tortoise coordinate r^* in (4.23) one shows that the time dependent perturbation is

$$\delta g_{x_1 x_2} = e^{-i\omega(t \mp r^*)} F(r^*, \omega)$$ (4.26)

Now, it is clear that the solution $\gamma = -\frac{i\omega}{f'(r_h)}$ corresponds to the infalling wave at the horizon. Therefore, one has

$$\phi \sim F(r_h, 0) \left[1 - \frac{i\omega}{f'(r_h)} \ln(r - r_h) + \ldots \right].$$ (4.27)

On the other hand, taking the near horizon limit of (4.20) one arrives at

$$\phi \sim c_1 + \frac{c_2 e^{-3A_h}}{f'(r_h)} \ln(r - r_h),$$ (4.28)

Comparison of (4.27) and (4.28) finally yields the coefficients $c_1 = F(r_h, 0)$ and $c_2 = -i\omega F(r_h, 0) e^{3A_h}$. Consequently using (4.12) and (4.19) one finds

$$\eta = \frac{1}{4\pi} \left(\frac{e^{3A_h}}{4G} \right).$$ (4.29)

We finally employ the Bekenstein-Hawking formula for black hole entropy (which equals the entropy of the dual plasma), $s = \frac{Area_{bh}}{4G}$, where numerator is area of the horizon. This then leads to our final result for the ratio of the shear viscosity to entropy in a generic holographic plasma:

$$\frac{\eta}{s} = \frac{1}{4\pi}.$$ (4.30)

We emphasize that the final answer is completely independent of the choice of the black hole background (4.16). Therefore, the shear viscosity to entropy ratio is *universal* and given by (4.30) for any large N plasma at strong coupling which is holographically dual to two-derivative Einstein's gravity coupled to arbitrary matter [19]. We also observe that the holographic value $1/4\pi \approx 0.08$ is consistent with the value on obtained from Bayesian analysis, Fig. 3.2. One does not find the T dependence of η/s in the holographic approach based on two-derivative gravity. Higher derivative corrections are needed to generate the temperature dependence, as was first described in [20]. We present the analogous holographic calculation of the bulk viscosity in

Chap. 9, Sect. 9.1, but before that we need to construct holographic theories dual to QCD-like non-conformal theories.

4.3 Holographic QCD

Main motivation for a realistic holographic dual of QCD is to understand the real-time dynamics in the *quark-gluon plasma* produced at the heavy ion collisions at RHIC and LHC (and perhaps in future planned FAIR and NICA). As we have seen above, holographic value for the shear viscosity to entropy ratio is much closer to the value inferred directly from ongoing experiments through Bayesian analyses than the value obtained from perturbative QCD calculations. We will argue below that this is not a coincidence, and holographic QCD indeed provides a faithful description of the many aspects of strongly interacting gauge theories. In the next chapters we will tailor a 5D holographic theory for QCD which describes most of the basic features of QCD (non-conformality, confinement, asymptotic, chiral symmetry breaking etc., see Sect. 2) and extend it to include the effects of finite magnetic fields.

4.3.1 What Do We Want from Holography?

The various observables we want to extract from the holographic description—in the order of increasing difficulty—are as follows:

- Spectrum of hadrons in the $T = 0$ ground state. As explained below, holographic QCD can capture at most spin-2 operators.[6] In this review we focus on the spectra of glueballs[7] and compare with available lattice QCD data. Computation of meson masses in a realistic holographic theory and comparison to the experimentally observed spectrum is an important problem which we will exclude from this review.[8]
- Thermodynamic functions. We will show that, in the realistic holographic theories, generically there exists a first order confinement-deconfinement transition at a finite temperature separating the confined state dual to the so-called "thermal gas" background and the deconfined state dual to a black hole. We then calculate the thermodynamic functions in Chap. 6, free energy, entropy and energy density as a function of T in the deconfined state, again comparing with available lattice QCD data.

---

[6] This shortcoming can be remedied by inclusion of higher spin fields but this increases the systematic ambiguities of holographic models [21].

[7] Baryon spectra in improved holographic QCD is an important open problem, see [22] for recent advances.

[8] See [23]—based on the holographic models we describe here—for the most recent progress.

- Hydrodynamic parameters. The zeroth order in thy hydrodynamics expansion is completely determined by the thermodynamics. At first order, there appears two non-trivial transport coefficients, the shear and the bulk viscosities, which we calculate in a realistic holographic model in Chap. 9 and again compare with available data. In addition, we include the so-called Chern-Simons decay rate, which is another transport coefficient that appears in the CP-odd sector of the theory and relevant for anomalous transport properties of the quark-gluon plasma introduced in Chap. 1 and detailed in Chap. 8.
- New observables that arise in the presence of external magnetic fields, B. We discuss these in Chap. 8. In particular we consider extension of the phase diagram in finite B, phenomena of (inverse) magnetic catalysis, and effects of magnetic fields static and transport coefficients such as the quark-anti-quark potential and the shear viscosity.

This is, of course, an incomplete list. A set of important observables relevant to QGP physics, among many others, concerns hard probes in heavy ion collisions, such as jet quenching, energy loss, momentum broadening and diffusion. For a holographic treatment of these observables see [7, 24]. Another involves non-equilibrium phenomena, especially relevant for the formation process of the quark-gluon plasma see e.g. [25–27].

4.3.2 Top-Down Versus Bottom-Up

QCD and other confining gauge theories are quite different than $\mathcal{N} = 4$ sYM theory and its deformations and holographic duality is less understood for these confining, non-conformal gauge theories. In general, there are two different approaches based on *critical* 10 dimensional string theory and *non-critical* 5 dimensional string theory. The first leads to what is referred to the *top-down* approach in the literature and the second leads to the *bottom-up* approach. Below, after briefly discussing the first, I will mostly focus on the latter approach.

Top-down approach: The top-down approach to holographic QCD starts from D-brane configurations in 10D string theory, such as N D4 branes wrapped on an S^1 in IIA string theory [28] and taking the so-called *decoupling limit* [29] which replaces the D-branes with a gravitational background including various matter fields. This approach[9] has later been generalized to include flavor degrees of freedom i.e. quarks in QCD by adding D8 "flavor" branes [30, 31]. This is a rigorous approach and, in principle, provides a precise dictionary between QFT on the D-branes and the dual gravity.

However, it is technically challenging because an accurate top-down holographic description of QCD requires solving the notorious problem of quantizing string theory in the presence of Ramond-Ramond form fields. In the case of conformal QFTs,

---

[9]There are many other top-down approaches in the literature which we do not discuss here.

such as $\mathcal{N} = 4$ sYM string modes can be parametrically separated from the low-lying supergravity modes (that are simply described by Einstein's gravity coupled to matter) by tuning the 't Hooft coupling λ to be large. This can be done because λ is a tunable parameter that does not run with the RG flow in a conformal theory. This option no longer exists in confining gauge theories such as QCD. The technical manifestation of this issue is that the supergravity limit is dual to QFT that contains infinitely many additional operators that arise from the Kaluza-Klein reduction from 10D to 5D on the extra 5 dimensional "internal" space of string background. If you try to make the KK-scale large (for example by sending the radius of S^1 to zero in the aforementioned D4 brane background) then stringy states become important. If you take the opposite limit by kicking string states out of the spectrums then there are infinitely many KK-states. This problem is generic to top-down theories and always shows up in different guises, [32]. This is essentially because the confinement energy scale, analogous Λ_{QCD} in QCD, mixes with the KK scale and the Hilbert space of these theories contain operators of arbitrarily large spin and scale dimensions, all proportional to Λ_{QCD}, with no parametric separation between low-lying operators and the rest.

Bottom-up approach: The origin of the technical difficulty in the top-down approach is the critical dimension of string theory in flat space, that is 10, which should be reduced to 5 (4 Minkowski field theory directions + 1 holographic direction). This procedure generates unwanted KK states. There exist however, and alternative type of string theory that exists in lower number of dimensions, called *non-critical* string theory. Non-critical strings [33] do not have Weyl-invariance on their world-sheet hence avoid the critical dimension constraint. In fact they can be mapped onto critical strings in one dimension higher but in non-flat background such as a dilaton field that linearly depends on one of the directions, which is known to satisfy the Weyl invariance on the worldsheet in a dimension lower than the critical 10D.[10] In fact, five dimensional non-critical string theory has been conjectured to be dual to pure Yang-Mills theory in 4D originally by Polyakov [8, 37, 38] even before the discovery of AdS/CFT correspondence [3]. In non-critical string theories, generically there is a non-trivial dilaton field which depends on one of the coordinates, and it is natural to identify the dual QFT operator as with the tr F^2 operator—that is the operator that couples to the gauge coupling constant in pure Yang-Mills—and the coordinates it depends on with the RG energy scale [8].

Even though holographic QCD based on non-critical string theory is (at least) as difficult as the top-down approach, it lends itself to an approximation by two-derivative Einstein's theory easier than the latter. It naturally suggests to consider Einstein's gravity coupled to a dilaton field Φ with some potential $V(\Phi)$. This potential should in principle be determined by the underlying non-critical string theory e.g. by solving the world-sheet beta function equations, but here one can use a short cut and determine $V(\Phi)$ by requirement of the fundamental properties of the dual field

---

[10] These theories typically display issues with instability and the bosonic non-critical string theory is ill defined for dimensions higher than D=2. Such issues can be side-stepped by considering the so-called 0B type strings [34–36].

theory: confinement, asymptotic freedom etc. This approach is called "bottom-up" holography. It provides a direct way to capture the IR dynamics of QCD in holography and became one of the main research directions in applied holography in mid 00s. It is hard to point to a single reference but some of the most notable papers developing this approach are [39–49]. The prize one pays is giving up a precise holographic dual to QCD, and be content with an effective theory capturing only the IR dynamics of relevant and/or marginal operators in the theory. As we will see below, bottom-up holography is quite useful and valuable in the absence of other approaches, especially for characterizing universal, qualitative features of strong coupling dynamics. Again, as we demonstrate below, it also provides a quantitatively good description for certain observables such as thermodynamic potentials.

Early bottom-up models [43, 44], sometimes called the "hard-wall" models consisted of an AdS_5 space terminating at a hard-wall at some location in the deep interior, to introduce the scale Λ_{QCD} and effectuate breaking of conformal symmetry. The main advantage of this model is its simplicity, calculations being almost identical to AdS. However, it leads to unrealistic results when applied to QCD such as vanishing trace anomaly, vanishing bulk viscosity, unrealistic behavior of thermodynamic functions in T, etc. It also leads to systematic errors in the hadron spectra arising from infinitely many possible boundary conditions one can impose at the hard cut-off (given by a mixture of Dirichlet and Neumann b.c.). Finally, the hard-wall model has the unrealistic feature of quadratic spectrum $m_n^2 \propto n^2$ for large excitation number n.

The "soft-wall" model was invented in [45] to overcome these difficulties. In these models the background consists of the AdS_5 metric and a dilaton field whose profile is chosen to mimic salient properties of QCD. The main purpose of [45] was to describe the meson spectra and interactions, which in this picture arise from space-filling (5D) "flavor" branes embedded in the geometry. The model indeed fulfils this purpose, however it leads to unrealistic features in the "glue" sector and in thermodynamics. See the short review [50] where a comparison of the "hard-wall", "soft-wall" and improved holographic models is provided. Almost all of these undesired aspects are removed by making the background dynamical. That is, instead of starting with a background designed by hand one obtains it by solving Einstein's gravity coupled to a scalar field. Below we explain the general construction of such dynamical soft-wall models [46–49] focusing on a specific model, called "improved holographic QCD" (ihQCD for short) [47, 48] which, in addition to the other salient features of QCD, also satisfies the correct UV behavior of the QCD coupling constant i.e. asymptotic freedom.

Improved holographic QCD: The operator product algebra and sum-rules [51] of QCD indicate that a sector of relevant and marginal operators can be treated separately from the rest of the Hilbert space of operators in the IR. This is motivated in the Chap. 1 by the Ward identities of QFT which form a closed subset of operators. We will expand this argument here. A basic operator which always remains in the Hilbert space, including the IR limit, is the stress tensor $T_{\mu\nu}$ that is conserved in the absence of external sources $\partial_\mu T^{\mu\nu} = 0$. An additional Ward identity comes from breaking of scale invariance by running couplings

$$T_\mu^\mu = \sum_i \beta_i(J)\langle \mathcal{O}_i \rangle , \qquad\qquad\qquad (4.31)$$

where the index i runs over all local gauge-invariant operators in the Hilbert space and β_i denote their associated beta-functions which are in principle functions of all sources J_i that couple to these operators. We will focus on pure Yang-Mills theory for simplicity.[11] In this case the only non-trivial source is the QCD coupling g itself which couples to the operator tr F^2. Therefore it is quite plausible that this subsector of the theory can indeed be described by Einstein's gravity coupled only to the dilaton field. The infinitely many scalar operators of Yang-Mills other than tr F^2 are irrelevant in the IR[12] and do not contribute to the dynamics. They can still have non-trivial VeVs in the ground state but these effects can be collectively included in the choice of the dilaton potential $V(\Phi)$ [52].

Now, our task is straightforward: construct an effective theory that correctly captures the physics that involves these low-lying operators in QCD using the basic ingredients from holography. For simplicity we first consider $SU(N)$ pure gauge theory in the large N limit and will add flavor degrees of flavor later. We already argued above that the minimal bottom-up holographic model should be Einstein's gravity coupled to a dilaton field with a non-trivial potential $V(\Phi)$ in this case. To determine the holographic relation between the 't Hooft coupling λ and the dilaton field Φ one adds a probe D3 brane in the geometry. Expanding its DBI action one finds $S_{D3} \propto \int d^4 x e^{-\Phi} F^2$ from which one determines the relation $\lambda \propto N \exp \Phi$. V should then be obtained by requiring the basic features of Yang-Mills/QCD.

- First of all, asymptotic freedom requires $\lambda \to 0$ in the UV. This translates into the statement $\Phi \to -\infty$ on the boundary of the 5D space-time. At the same time the UV of the field theory is trivially scale invariant given by a Gaussian fixed point. Such an asymptotically free gauge theory is also expected to be dual to an asymptotically AdS geometry.[13]
  Hence we expect $V(\Phi) \to 12/\ell^2$ as $\Phi \to -\infty$, where $-12/\ell^2$ is the cosmological constant that leads to the AdS$_5$ geometry. This constant term is corrected by subleading terms

$$V(\Phi) = \frac{12}{\ell^2} + v_1 e^\Phi + v_2 e^{2\Phi} + \cdots , \qquad \Phi \to -\infty , \qquad (4.32)$$

  whose form are determined by running of the 't Hooft coupling. This expansion is derived in the next chapter. In particular the coefficients v_i are fixed in terms of the beta-function coefficients of large N pure Yang-Mills (or large N QCD once

---

[11] We include fundamental representations later in Chap. 7.

[12] In addition there is the topological charge operator tr $F \wedge F$ but, as we discuss in Chap. 8, this operator is suppressed by $1/N$ in the 't Hooft limit hence can be treated as a perturbation on the background of the first two operators.

[13] Demanding asymptotic freedom in holographic QCD should be understood as fixing the UV boundary conditions at a cut-off scale beyond which holographic description fails. Holographic duals of weakly coupled theories generally involve string corrections to all orders [10].

flavor is included). The leading constant term is indeed generated from curvature corrections in the effective string action [47]. This is our first requirement on the dilaton potential $V(\Phi)$.

- Second requirement is linear confinement of quarks, that quark anti-quark potential behaves as $V_{q\bar{q}} = \sigma_0 L + \cdots$ for $L \gg 1$ where L is the distance between test charges. In the holographic dual, test quarks are realized as end-points of open strings on the boundary. Therefore linear confinement translates into linearity of the Nambu-Goto action of this probe string for a large separation L between endpoints. As we discuss in the next chapter, this requirement restricts the large Φ, IR behavior of the dilaton potential to be of the form:

$$V(\Phi) \propto e^{\frac{4}{3}\Phi} \Phi^P, \quad P > 0, \quad \text{or} \quad V(\Phi) \propto e^{Q\Phi}, \; Q > 4/3 \quad \Phi \gg 1. \quad (4.33)$$

- There will be the additional requirement from chiral symmetry breaking (see Chap. 2) on the gravitational potentials once we include flavor in the holographic description in Chap. 7.

We emphasize that implementation of both asymptotic freedom and linear confinement has always been the holy grail of the gauge-string duality [24]. The theory we outlined above, improved holographic QCD, provides an alternative viewpoint on this long standing open problem. On the other hand, we should also emphasize that this construction is not based on a first-principles derivation but only plausibility arguments. It remains open how to precisely embed this construction in 5D non-critical string theory and how non-critical string theory precisely maps to QCD [8].

References

1. J. Polchinski, Phys. Rev. Lett. **75**, 4724–4727 (1995). https://doi.org/10.1103/PhysRevLett.75. 4724 [arXiv:hep-th/9510017 [hep-th]]
2. J. Polchinski, [arXiv:hep-th/9611050 [hep-th]]
3. J.M. Maldacena, Int. J. Theor. Phys. **38**, 1113 (1999) [Adv. Theor. Math. Phys. **2**, 231 (1998)]. https://doi.org/10.1023/A:1026654312961 [hep-th/9711200]
4. S.S. Gubser, I.R. Klebanov, A.M. Polyakov, Phys. Lett. B **428**, 105–114 (1998). https://doi. org/10.1016/S0370-2693(98)00377-3 [arXiv:hep-th/9802109 [hep-th]]
5. E. Witten, Adv. Theor. Math. Phys. **2**, 253–291 (1998). https://doi.org/10.4310/ATMP.1998. v2.n2.a2 [arXiv:hep-th/9802150 [hep-th]]
6. G. 't Hooft, Conf. Proc. C **930308**, 284–296 (1993) [arXiv:gr-qc/9310026 [gr-qc]]
7. L. Susskind, J. Math. Phys. **36**, 6377–6396 (1995). https://doi.org/10.1063/1.531249 [arXiv:hep-th/9409089 [hep-th]]
8. A.M. Polyakov, Int. J. Mod. Phys. A **14**, 645–658 (1999). https://doi.org/10.1142/ S0217751X99000324 [arXiv:hep-th/9809057 [hep-th]]
9. A.M. Polyakov, Int. J. Mod. Phys. A **17S1**, 119–136 (2002). https://doi.org/10.1142/ S0217751X02013071 [arXiv:hep-th/0110196 [hep-th]]
10. R. Gopakumar, Phys. Rev. D **70**, 025009 (2004). https://doi.org/10.1103/PhysRevD.70.025009 [arXiv:hep-th/0308184 [hep-th]]

11. S. Ryu, T. Takayanagi, Phys. Rev. Lett. **96**, 181602 (2006). https://doi.org/10.1103/PhysRevLett.96.181602 [arXiv:hep-th/0603001 [hep-th]]
12. M. Van Raamsdonk, Gen. Rel. Grav. **42**, 2323–2329 (2010). https://doi.org/10.1142/S0218271810018529 [arXiv:1005.3035 [hep-th]]
13. G. Penington, JHEP **09**, 002 (2020). https://doi.org/10.1007/JHEP09(2020)002 [arXiv:1905.08255 [hep-th]]
14. A. Almheiri, N. Engelhardt, D. Marolf, H. Maxfield, JHEP **12**, 063 (2019). https://doi.org/10.1007/JHEP12(2019)063 [arXiv:1905.08762 [hep-th]]
15. G. 't Hooft, Nucl. Phys. B **72**, 461 (1974). https://doi.org/10.1016/0550-3213(74)90154-0
16. E. D'Hoker, D.Z. Freedman, [arXiv:hep-th/0201253 [hep-th]]
17. G. Policastro, D.T. Son, A.O. Starinets, Phys. Rev. Lett. **87**, 081601 (2001). https://doi.org/10.1103/PhysRevLett.87.081601 [arXiv:hep-th/0104066 [hep-th]]
18. S.S. Gubser, A. Nellore, S.S. Pufu, F.D. Rocha, Phys. Rev. Lett. **101**, 131601 (2008). https://doi.org/10.1103/PhysRevLett.101.131601 [arXiv:0804.1950 [hep-th]]
19. A. Buchel, J.T. Liu, Phys. Rev. Lett. **93**, 090602 (2004). https://doi.org/10.1103/PhysRevLett.93.090602 [arXiv:hep-th/0311175 [hep-th]]
20. S. Cremonini, U. Gursoy, P. Szepietowski, JHEP **08**, 167 (2012). https://doi.org/10.1007/JHEP08(2012)167 [arXiv:1206.3581 [hep-th]]
21. A. Karch, E. Katz, D.T. Son, M.A. Stephanov, Phys. Rev. D **74**, 015005 (2006). https://doi.org/10.1103/PhysRevD.74.015005 [arXiv:hep-ph/0602229 [hep-ph]]
22. T. Ishii, M. Järvinen, G. Nijs, JHEP **07**, 003 (2019). https://doi.org/10.1007/JHEP07(2019)003 [arXiv:1903.06169 [hep-ph]]
23. A. Amorim, M.S. Costa, M. Järvinen, [arXiv:2102.11296 [hep-ph]]
24. U. Gursoy, "Improved Holographic QCD and the Quark-gluon Plasma," Acta Phys. Polon. B **47**, 2509 (2016). https://doi.org/10.5506/APhysPolB.47.2509
25. P.M. Chesler, L.G. Yaffe, Phys. Rev. Lett. **102**, 211601 (2009). https://doi.org/10.1103/PhysRevLett.102.211601 [arXiv:0812.2053 [hep-th]]
26. M.P. Heller, R.A. Janik, P. Witaszczyk, Phys. Rev. Lett. **108**, 201602 (2012). https://doi.org/10.1103/PhysRevLett.108.201602 [arXiv:1103.3452 [hep-th]]
27. J. Casalderrey-Solana, M.P. Heller, D. Mateos, W. van der Schee, Phys. Rev. Lett. **111**, 181601 (2013). https://doi.org/10.1103/PhysRevLett.111.181601 [arXiv:1305.4919 [hep-th]]
28. E. Witten, Adv. Theor. Math. Phys. **2**, 505–532 (1998). https://doi.org/10.4310/ATMP.1998.v2.n3.a3 [arXiv:hep-th/9803131 [hep-th]]
29. O. Aharony, S.S. Gubser, J.M. Maldacena, H. Ooguri, Y. Oz, Phys. Rept. **323**, 183–386 (2000). https://doi.org/10.1016/S0370-1573(99)00083-6 [arXiv:hep-th/9905111 [hep-th]]
30. T. Sakai, S. Sugimoto, Prog. Theor. Phys. **113**, 843–882 (2005). https://doi.org/10.1143/PTP.113.843 [arXiv:hep-th/0412141 [hep-th]]
31. T. Sakai, S. Sugimoto, Prog. Theor. Phys. **114**, 1083–1118 (2005). https://doi.org/10.1143/PTP.114.1083 [arXiv:hep-th/0507073 [hep-th]]
32. U. Gursoy, JHEP **05**, 014 (2006). https://doi.org/10.1088/1126-6708/2006/05/014 [arXiv:hep-th/0602215 [hep-th]]
33. J. Polchinski, https://doi.org/10.1017/CBO9780511816079
34. I.R. Klebanov, A.A. Tseytlin, Nucl. Phys. B **546**, 155–181 (1999). https://doi.org/10.1016/S0550-3213(99)00041-3 [arXiv:hep-th/9811035 [hep-th]]
35. I.R. Klebanov, A.A. Tseytlin, Nucl. Phys. B **547**, 143–156 (1999). https://doi.org/10.1016/S0550-3213(99)00084-X [arXiv:hep-th/9812089 [hep-th]]
36. E. Alvarez, C. Gomez, Nucl. Phys. B **550**, 169–182 (1999). https://doi.org/10.1016/S0550-3213(99)00142-X [arXiv:hep-th/9902012 [hep-th]]
37. A.M. Polyakov, Nucl. Phys. B **486**, 23–33 (1997). https://doi.org/10.1016/S0550-3213(96)00601-3 [arXiv:hep-th/9607049 [hep-th]]
38. A.M. Polyakov, Nucl. Phys. B Proc. Suppl. **68**, 1–8 (1998). https://doi.org/10.1016/S0920-5632(98)00135-2 [arXiv:hep-th/9711002 [hep-th]]
39. J. Polchinski, M.J. Strassler, Phys. Rev. Lett. **88**, 031601 (2002). https://doi.org/10.1103/PhysRevLett.88.031601 [hep-th/0109174]

40. S. Kuperstein, J. Sonnenschein, JHEP **0407**, 049 (2004). https://doi.org/10.1088/1126-6708/2004/07/049 [hep-th/0403254]
41. I.R. Klebanov, J.M. Maldacena, Int. J. Mod. Phys. A **19**, 5003 (2004). https://doi.org/10.1142/S0217751X04020865 [hep-th/0409133]
42. F. Bigazzi, R. Casero, A.L. Cotrone, E. Kiritsis, A. Paredes, JHEP **0510**, 012 (2005). https://doi.org/10.1088/1126-6708/2005/10/012 [hep-th/0505140]
43. J. Erlich, E. Katz, D.T. Son, M.A. Stephanov, Phys. Rev. Lett. **95**, 261602 (2005). https://doi.org/10.1103/PhysRevLett.95.261602 [hep-ph/0501128]
44. L. Da Rold, A. Pomarol, Nucl. Phys. B **721**, 79 (2005). https://doi.org/10.1016/j.nuclphysb.2005.05.009 [hep-ph/0501218]
45. A. Karch, E. Katz, D.T. Son, M.A. Stephanov, Phys. Rev. D **74**, 015005 (2006). https://doi.org/10.1103/PhysRevD.74.015005 [hep-ph/0602229]
46. C. Csaki, M. Reece, JHEP **0705**, 062 (2007). https://doi.org/10.1088/1126-6708/2007/05/062 [hep-ph/0608266]
47. U. Gursoy, E. Kiritsis, JHEP **0802**, 032 (2008). https://doi.org/10.1088/1126-6708/2008/02/032 [arXiv:0707.1324 [hep-th]]
48. U. Gursoy, E. Kiritsis, F. Nitti, JHEP **0802**, 019 (2008). https://doi.org/10.1088/1126-6708/2008/02/019 [arXiv:0707.1349 [hep-th]]
49. S.S. Gubser, A. Nellore, Phys. Rev. D **78**, 086007 (2008). https://doi.org/10.1103/PhysRevD.78.086007 [arXiv:0804.0434 [hep-th]]
50. U. Gursoy, Mod. Phys. Lett. A **23**, 3349 (2009). https://doi.org/10.1142/S0217732308029940 [arXiv:0904.2750 [hep-th]]
51. M.A. Shifman, A.I. Vainshtein, V.I. Zakharov, Nucl. Phys. B **147**, 385 (1979). https://doi.org/10.1016/0550-3213(79)90022-1
52. E. Kiritsis, Fortsch. Phys. **57**, 396–417 (2009). https://doi.org/10.1002/prop.200900011 [arXiv:0901.1772 [hep-th]]

Chapter 5
Improved Holographic QCD

5.1 Construction of the Theory

As outlined in the previous chapter, our starting point for the bottom-up holographic dual to pure SU(N) Yang-Mills theory in the large N limit is the Einstein-dilaton action[1]:

$$S = M_p^3 N^2 \int \sqrt{-g}\, d^5 \left(R - \frac{4}{3}(\partial\Phi)^2 + V(\Phi) \right) + \mathrm{GH} + S_{ct} , \qquad (5.1)$$

where M_p is the Planck energy of the 5D theory (a parameter of the model which we fix below) related to 5D Newton's constant as $M_p^3 N^2 = 1/16\pi G$ and the N dependence in front of the action is made explicit,[2] to match the same scaling of pure Yang-Mills action at large N. GH denotes the Gibbons-Hawking term included to render the variational problem of the metric well-defined in geometries with boundary, and the last term is the standard holographic counter-term to make the on-shell action finite on asymptotically AdS spacetimes [2, 3]. The GH term is given by

$$\mathcal{S}_{GH} = 2M_p^3 \int_{\partial M} d^4x \sqrt{h}\, K \qquad (5.2)$$

with

$$K_{\mu\nu} \equiv -\nabla_\mu n_\nu = \frac{1}{2} n^\rho \partial_\rho h_{\mu\nu} \quad , \quad K = h^{ab} K_{ab} \qquad (5.3)$$

---

[1] The unconventional normalization of the dilaton kinetic term follows from the underlying non-critical string theory in 5D [1]. Conventional normalization can be attained by rescaling $\Phi \to \sqrt{3/8}\Phi$.

[2] This is done by rescaling the original dilaton in the string frame as $\exp -\Phi \to N \exp -\Phi$ and Weyl transforming the metric to Einstein frame by $g = \exp(4\Phi/3)g_E$ [1].

© The Author(s), under exclusive license to Springer Nature Switzerland AG 2021
U. Gürsoy, *Holography and Magnetically Induced Phenomena in QCD*,
SpringerBriefs in Physics,
https://doi.org/10.1007/978-3-030-79599-3_5

where h_{ab} is the induced metric on the boundary and n_μ is the (outward) unit vector normal to the boundary. The potential V in (5.1) is assumed to include a negative cosmological constant, which, in the absence of Φ, assures presence of an AdS solution.

In the *vacuum state*, at vanishing temperature the boundary theory enjoys Lorentz symmetry $SO(3, 1)$, which should be reflected in the isometries of the corresponding gravity solution. Therefore both the dilaton and the metric functions will be assumed to depend only on the holographic coordinate r. Hence the following ansatz for the metric can be made with no loss of generality

$$ds^2 = e^{2A(r)} \left(dr^2 + \eta_{\mu\nu} dx^\mu dx^\nu \right) . \tag{5.4}$$

In this coordinate frame, which we call *conformal coordinates*, r runs from the boundary at $r = 0$ to the deep interior which may terminate at $r = const.$ or $r = \infty$ depending on the potential V. Einstein's equations then reduce to

$$\ddot{A} - (\dot{A})^2 = -\frac{4}{9}(\dot{\Phi})^2, \qquad 3\ddot{A} + 9(\dot{A})^2 = e^{2A} V(\Phi), \tag{5.5}$$

where dot denotes derivative with respect to r. The equation of motion of the dilaton is not independent and can be derived from these two equations [1].

It is also useful to work with another coordinate frame which we call the *domain-wall coordinates* that follows from (5.4) by substituting $dr = e^{-A} du$

$$ds^2 = du^2 + e^{2A(u)} \eta_{\mu\nu} dx^\mu dx^\nu . \tag{5.6}$$

Here u runs from the boundary at $u = -\infty$ to the deep interior which terminates at $u = u_0$. Einstein's equations in this coordinate system are

$$A'' = -\frac{4}{9}(\Phi')^2, \qquad 3A'' + 12A'^2 = V(\Phi), \tag{5.7}$$

where prime denotes derivative with respect to u.

5.2   UV Asymptotics

We first consider the near boundary asymptotics of the gravitational background which corresponds to UV of the dual QFT. We require the metric asymptote to AdS near the boundary:

$$A(u) \to -u/\ell + \cdots , \qquad u \to -\infty , \tag{5.8}$$

where ℓ is the AdS radius related to 5D cosmological constant as $-12/\ell^2$. We note that the first equation in (5.5) requires that the derivative A' is monotonically decreasing. This fact can be traced back to the null-energy condition in the 5D space-time and directly related to the c-theorem in the dual QFT [4]. Another important piece of information which follows from the requirement of asymptotic AdS is the fact that [5] $A' = -1/\ell < 0$ as $u \to -\infty$. Combining this with monotonicity of A' (see (5.7)), one finds that $A(u) \to -\infty$ at some point $u = u_0$. At this point there is a curvature singularity as one can directly verify by computing the Ricci scalar [1]. Such possible singularities were classified in [5]. Crucially, a particular class of such singularities [5], called *repulsive*, are acceptable in the context of holography [6] as we explain below.

The second equation in (5.7) requires $V \to 12/\ell^2$ on the boundary. This is minus the value of the cosmological constant corresponding to AdS_5 space-time and it constitutes the leading term of the dilaton potential in the UV limit. Now we determine the subleading terms in this limit, recalling that, Φ corresponds to the operator tr F^2 in Yang-Mills theory. In the UV limit where the theory becomes free, its scaling dimension goes to $\Delta = 4$. Quantum loops close to the UV make it slightly less than the marginal value 4.

This leaves us with two options:

1. One declares that the boundary field theory in the UV is a strongly interacting CFT deformed by a relevant operator $\mathcal{O} = \text{tr } F^2$ with scaling dimension close to but smaller than 4, $\Delta = 4 - \epsilon$. Then the corresponding field has a small mass given by the usual AdS/CFT formula (4.6). In this case the potential has a true minimum

$$V = \frac{12}{\ell^2} + \frac{m^2}{2}(\Phi - \Phi_0)^2 + \cdots \qquad (5.9)$$

and the UV fixed point corresponds to the value $\Phi = \Phi_0$. This choice is advocated in [7, 8] and has the advantage of being more familiar in the AdS/CFT context precisely because it is obtained by deforming a 4D CFT.[3] In particular, the procedure of holographic renormalization is simple [2]. These examples are easier to embed in top-down string theory rather than non-critical string. String corrections are controlled by the 't Hooft coupling, see (4.3), (4.4). However, this choice does not correspond to actual QCD behavior which is asymptotically free, $\Delta \to 4$ in the UV precisely, and the coupling constant never stops running. It also may also have other disadvantages as the corresponding vacuum and thermal states may be unstable [11–13].

2. One takes $\Delta = 4$ exactly in the UV. In this case, the dilaton field is massless and the UV asymptotics of the dilaton potential will be qualitatively different, exhibiting a runaway behavior rather than a true minimum. This case mimics the running of the coupling constant and the dimension Δ in QCD. The string

---

[3] Non-trivial non-supersymmetric CFTs in 4D may exist [9]. A clear example follows from orbifolding $\mathcal{N} = 4$ super Yang-Mills [10]. Whether these are useful to model the UV of QCD is unclear to us.

corrections are controlled by the Ricci scalar as usual, but the relation (4.3) that relates this to the 't Hooft coupling is not any more valid for these solutions. Instead, the coupling is free to run and vanish in UV. We will work with this choice and call it *improved holographic QCD* [1]. Below, we explain how to fix the UV asymptotics of the potential using the known beta-function of pure SU(N) theory. The main disadvantage of this choice is that holographic renormalization is non-standard and more complicated. Yet, it is well-defined and worked out in detail in [3].

The perturbative beta-function of pure SU(N) YM gauge theory in the large N limit is given by

$$\beta(\lambda) = \frac{d\lambda}{d\ln E} = -b_0\lambda^2 - b_1\lambda^3 + \cdots \tag{5.10}$$

in the limit $\lambda \ll 1$, i.e. in the UV. Here the first two beta-function coefficients

$$b_0 = \frac{22}{3(4\pi)^2}, \qquad b_1 = \frac{51}{121}b_0^2, \tag{5.11}$$

are scheme-independent and positive definite leading to asymptotic freedom. The higher order coefficients are scheme-dependent as can be shown by a redefinition of λ. Now we want to connect this UV behavior to holography near the boundary. It turns out that the holographic theory can not be trusted in the far UV because the higher derivative corrections to gravity—which we neglect in the two-derivative Einstein theory—become important. One can easily check that the Ricci curvature in the string frame[4] behaves like $R_s \sim (-\log(r/\ell))^{\frac{4}{3}}$ as $r \to 0$. Indeed we do not trust the theory in the far UV limit, however we may still use the identification with running of the perturbative QCD theory to provide *initial conditions* for the holographic RG flow. The initial conditions set at small λ determine the behavior of the theory at intermediate and strong λ, that is, in the IR, the regime well approximated by the holographic description.

The next question is how λ and its RG flow relate to the corresponding quantities in the dual gravitational theory. As mentioned above, the dilaton, more precisely $\exp\Phi$ couples to the operator $\mathrm{tr}\,F^2$ on a probe D3 brane in the gravitational background [14], hence its non-normalizable and normalizable modes, see Eq. (4.9), are associated with the 't Hooft coupling and the VeV $\langle \mathrm{tr}\,F^2 \rangle$ respectively. On the other hand the energy scale E is related to the conformal factor scale $\exp A$ in the metric (5.4) [15]. The motivation for this identification comes from the fact that the energy of a state at location u in the interior of the geometry, measured by an asymptotic observer on the

---

[4] String frame, which is related to the Einstein frame as $g_{\mu\nu}^s = \exp(4\Phi/3)g_{\mu\nu}^E$, is the relevant frame to compute the higher derivative corrections. The Ricci scalars in the two frames are related as $R_s = \exp(-4\Phi/3)(R_E - 16\nabla^2\Phi/3 - 16/3\partial_\mu\Phi\partial^\mu\Phi)$.

boundary involves the factor exp A because of the gravitational red-shift measured by g_{tt} [14]. This motivates the identifications[5]

$$\lambda = \exp \Phi(u), \qquad \ln E = A(u). \tag{5.12}$$

The second choice corresponds a particular holographic renormalization scheme, see [1] for a discussion of scheme dependences in these identifications. With these identifications one finds,

$$\beta(\lambda) = \frac{d\lambda}{dA}. \tag{5.13}$$

This gravitational quantity corresponds to the beta-function in the dual QFT in what we called holographic renormalization scheme (5.12). We will now show that this quantity can be directly determined by solving Einstein's equations. One can, in fact, derive a "master equation" for the holographic beta-function. For technical reasons it is slightly easier to work with the following *scalar variable* [17]

$$X(\Phi) \equiv \frac{\beta(\lambda)}{3\lambda} = \frac{d\Phi}{3dA} = \frac{1}{3}\frac{\Phi'(u)}{A'(u)}. \tag{5.14}$$

It is related to the well-known "fake superpotential" $W(\Phi)$ in the gravitational theory by $X = -3/4 \, d\ln(W)/d\Phi$ [17, 18]. One can easily derive, see Appendix A, the equation of motion for the scalar variable X defined above, starting from Einstein's equations (6.9). One finds

$$\frac{dX}{d\Phi} = -\frac{4}{3}(1 - X^2)\left(1 + \frac{3}{8}\frac{1}{X}\frac{d\log V}{d\Phi}\right). \tag{5.15}$$

From this one obtains the following master equation for the beta function:

$$\frac{d\beta}{d\lambda} = \frac{\beta}{\lambda} - 4\left(1 - \frac{\beta^2}{9\lambda^2}\right)\left(1 + \frac{9}{8}\frac{\lambda^2}{\beta}\frac{d\log V}{d\lambda}\right). \tag{5.16}$$

We will see below that there is a single consistent boundary condition for this first-order differential equation which should be imposed at the curvature singularity located at $\lambda = \infty$. This is the condition for an acceptable singularity, i.e. the condition that the curvature singularity can be cloaked by a tiny horizon [6]. The condition is,

$$\lim_{\lambda \to \infty} \frac{\beta(\lambda)}{\lambda^2 \frac{d\log V}{d\lambda}} = -\frac{9}{8}. \tag{5.17}$$

---

[5] There is the possibility of including a constant multiplicative factor in the first identification [16] which corresponds to RG scheme dependence in the field theory. We set this multiplicative factor to 1 in these notes.

Solving (5.16) with the condition (5.17) *uniquely* determines the beta-function of the dual QFT hence *the information contained in the dilaton potential V is in one-to-one correspondence with the beta function of the QFT.*

We will assume that the solution of this equation is negative definite throughout the entire range $0 \leq \lambda \leq \infty$:

$$\beta(\lambda) < 0 . \tag{5.18}$$

This corresponds to the assumption that there is no IR fixed point in the theories we want to consider. We learn from the definition (5.14) that $\Phi' > 0$, since $A' < 0$ as we explained above. Consistently, we will assume that the coupling constant in the dual field theory grows indefinitely towards the IR. Hence the dilaton diverges at the deep interior of the geometry:

$$\phi(u) \to \infty, \qquad u \to u_0 . \tag{5.19}$$

As we show below, the curvature scalar in the string frame vanishes for the backgrounds we consider, thus (5.19) does not force us to include corrections from the higher string states. It does, however, makes questionable the absence of loop corrections to non-critical string theory which we ignored so far. These corrections are proportional to the string coupling constant g_s which is itself proportional to $\exp(\phi)$. Recalling that the proportionality constant between the actual string dilaton and $\exp(\phi)$ is $1/N$, see below Eq. (5.1), string interactions can indeed be neglected as long as one takes the strict $N \to \infty$ limit first. This is an important restriction of the holographic approach to QCD: one should really work in the strict $N = \infty$ limit in order to avoid string loop corrections. Luckily, there are good indications, see Fig. 6.3, that $N = 3$ and $N = \infty$ are not too far from each other, as long as one judiciously normalizes the observables by a factor of N^2.

Through Eqs. (5.10), (5.13) and (5.15) one obtains the desired UV expansion of the dilaton potential as,

$$V(\Phi) = \frac{12}{\ell^2} \left(1 + v_0 e^\Phi + v_1 e^{2\Phi} + \cdots\right) \qquad v_0 = \frac{8}{9} b_0, \ v_1 = \frac{1}{81} \left(23b_0^2 + 36b_1\right) . \tag{5.20}$$

This determines the UV asymptotics of the ihQCD potential. Given (5.20) one obtains the near boundary asymptotics of the background by solving (5.5). In the conformal coordinate frame the boundary is at $r = 0$ and the expansion of the background reads

$$e^{A(r)} = \frac{\ell}{r} \left[1 + \frac{4}{9} \frac{1}{\ln[r\Lambda]} - \frac{4b_1}{9b_0^2} \frac{\ln[-\ln[r\Lambda]]}{\ln^2[r\Lambda]} + \cdots\right] , \tag{5.21}$$

$$b_0 e^{\Phi(r)} = -\frac{1}{\ln[r\Lambda]} + \frac{b_1}{b_0^2} \frac{\ln[-\ln[r\Lambda]]}{\ln^2[r\Lambda]} + \cdots \tag{5.22}$$

Here Λ is an integration constant, associated to the running of the coupling in the dual field theory and will be identified with the IR scale Λ_{QCD}.

5.3 IR Asymptotics

IR asymptotics of the dilaton potential is determined by the requirement of quark confinement. In QCD-like confining theories the potential between a test quark and a test anti-quark goes linearly in the separation L for large L

$$V_{q\bar{q}}(L) = \sigma_0 L + \cdots , \qquad L \gg 1/\Lambda , \qquad (5.23)$$

Here σ_0 is the QCD string tension. As explained above, linear quark confinement can be qualitatively understood in terms of a gluon flux tube connecting the quark and the anti-quark, see Fig. 2.1.

This quark-anti-quark potential is dual on the gravity side to the action of a string with endpoints at the locations $x = 0$ and $x = L$ [19, 20]:

$$t V_{q\bar{q}}(L) = S_{NG} + S_d - S_{ct} = \frac{1}{2\pi \ell_s^2} \int_0^t d\tau d\sigma \sqrt{-\det h_{\alpha\beta}} + S_d - S_{ct} , \quad (5.24)$$

where we denote the space-time coordinates by X^μ and we have chosen the gauge $X^0 = \tau$. ℓ_s is the string length scale. One also typically chooses $\sigma = X^1 = x$. The world-sheet metric is $h_{\alpha\beta} = \partial_\alpha X^\mu \partial_\beta X^\nu g_{\mu\nu}^s$, where G^s is the background metric in the *string frame*, related to the metric (5.4) in the Einstein frame as:

$$ds_{st}^2 = e^{2A_s(r)}(dr^2 + \eta_{\mu\nu} dx^\mu dx^\nu), \qquad A_s(r) = A(r) + \frac{2}{3}\Phi(r) . \quad (5.25)$$

We included a counter-term S_{ct} in (5.24) because the on-shell string action diverges on asymptotically AdS space-times. There is a standard way to determine this counter-term action [21] and the detailed calculation for the ihQCD backgrounds is presented in [5]. One should also recall that, in backgrounds with a non-trivial dilaton profile there is an additional term

$$S_s = \int d\sigma \delta\tau \sqrt{-g} R^{(2)} \Phi(X^\mu) , \quad (5.26)$$

where $R^{(2)}$ is the world-sheet Ricci scalar. This term is typically topological, counting handles on the closed string, but not when dilaton has dependence on the target spacetime coordinates the string is embedded. This term is computed in Appendix C of [5] and shown that it does not modify the qualitative results discussed below.

The generic mechanism that generates behavior (5.23) from (5.24) is as follows [22]: when the geometry ends at a specific point $r = r_0$ deep in the interior then the tip of the string hanging from the boundary to the interior will get stuck at this locus to minimize its energy. As one takes the end points further apart in the limit $L \to \infty$ then there will be a contribution from this tip proportional to L. This is how the hard-wall background of [23, 24] confines quarks: the tip of the string gets stuck at the location of the hard-wall, since the geometry ends there.

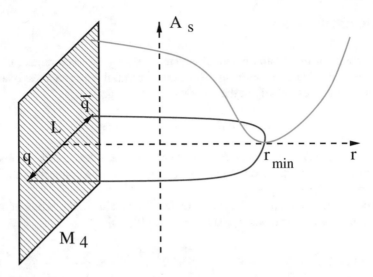

Fig. 5.1 Linear confinement is generated by minimum of the string-frame scale factor in holographic QCD theories

This mechanism is generalized in [5] where it is shown that, in the large K limit the tip of the string will generally be located at the minimum of the string-frame scale factor e^{2A_s}, $r = r_{min}$. Therefore, the action becomes proportional to L in the large L limit, whenever the string-frame scale factor has a minimum. This is pictorially described in Fig. 5.1. A simple calculation [5] shows that the QCD string tension σ_0 in (2.5) is related to the tension of the string in 5D as

$$\sigma_0 = \frac{e^{2A_s(r_{min})}}{2\pi \ell_s^2} \, . \tag{5.27}$$

The mechanism outlined above is the most general one leading to linear quark confinement and finite QCD string tension [5].

The next question is how this requirement translates into a condition on the dilaton potential. From the UV asymptotics in Sect. 5.2 it is clear that the string frame scale factor $A_s = A + 2\Phi/3$ in (5.25) diverges on the boundary and starts decreasing from the boundary towards the interior. In order to acquire a minimum at r_{min} it should start increasing at an intermediate locus in r. Assuming for simplicity a single minimum of the function A_s, this requirement translates into divergence of A_s at the IR end point of the geometry, which we denote by r_0 in the conformal coordinate system[6] where $r_0 > r_{min}$. For this to happen, as can be seen clearly from (5.25), we have to require

$$\frac{dA}{d\Phi} > -\frac{2}{3} \rightarrow X < -\frac{1}{2} \, . \tag{5.28}$$

---

[6] This corresponds to u_0 in domain-wall coordinates.

as $r \to r_0$. A more careful analysis [5] shows that

$$\lim_{\Phi \to \infty} \left(X + \frac{1}{2} \right) \Phi = K, \qquad 0 \geq K \geq -\infty. \tag{5.29}$$

This means that for linear confinement to take place the scalar variable X should approach $-1/2$ from below with the rate K/Φ.

From (5.15) it is clear that this can only happen when X hits one of the zeros of the RHS as $\Phi \to \infty$:

$$\text{I.} \qquad X \to -\frac{3}{8} \frac{V'(\Phi)}{V(\Phi)} \bigg|_{\Phi = \infty}, \tag{5.30}$$

$$\text{II.} \qquad X \to -1, \tag{5.31}$$

$$\text{III.} \qquad X \to +1. \tag{5.32}$$

It is immediate to rule out case II and III. In case II, the curvature singularity at $\Phi = \infty$ is not of the good type according to the criteria in [6]. In particular it is not possible to cloak the singularity behind an infinitesimal horizon. Case III requires X passing from 0 which violates our assumption (5.18). In particular, the β function goes through a zero that would correspond to a local fixed point.[7]

On the other hand, case I arises only for a specific dilaton potential, IR (large Φ) regime of which is dominated by an exponential term of the form $V \to \exp(2Q\Phi)\Phi^P$ up to subleading corrections, where both $Q > 0$ and P real numbers. This is because only then the RHS of (5.30) is finite. We write this using the notation introduced in (5.29) as

$$V(\Phi) \to V_\infty e^{\frac{4}{3}\Phi} \Phi^{-\frac{8}{3}K}, \qquad \Phi \to \infty, \tag{5.33}$$

with where V_∞ is some constant. The IR background geometry is determined by the constant K [5]. The particular case $K = -\infty$ corresponds to purely exponential. In this case the asymptotics of the potential should be chosen as,

$$V(\Phi) \to V_\infty e^{-\frac{8}{3}X(\infty)\Phi}, \qquad \Phi \to \infty, \tag{5.34}$$

with $X(\infty) < -1/2$ which translates into the second option in Eq.[8] (4.33). We will leave out the special case of $K = -\infty$ out in our analysis below.

Solving Einstein's equations with the requirement (5.33) one finds [5] that there are three classes of confining IR geometries:

---

[7] Even though this is not useful for QCD applications, it contains interesting physics, coined "exotic RG flows" and have recently been studied in [11–13].

[8] One sees from (5.15) that $X = -1$ is an attractive fixed point and $X(\infty)$ cannot be smaller than this value [1]. A more strict condition on this exponent comes from analyzing glueball spectrum [5].

$$A \to -Cr^{\alpha}, \qquad -\frac{3}{8} < K \le 0, \qquad K \equiv -\frac{3}{8}\frac{\alpha - 1}{\alpha} \tag{5.35}$$

$$A \to -C(r_0 - r)^{-\tilde{\alpha}}, \qquad -\infty < K < -\frac{3}{8}, \qquad K \equiv -\frac{3}{8}\frac{\tilde{\alpha} + 1}{\tilde{\alpha}} \tag{5.36}$$

$$A \to \delta \log(r_0 - r), \qquad X(\infty) = \frac{2}{3}\sqrt{1 + 1/\delta} < -\frac{1}{2}, \qquad K = -\infty \tag{5.37}$$

where C is an integration constant determined in terms of Λ. In particular the metric has a curvature singularity at $r = r_0 < \infty$ when $K < -3/8$ and $r_0 = \infty$ when $0 \ge K > -3/8$. The dilaton behaves as

$$\Phi(r) \to -\frac{3}{2}A(r) + \frac{3}{4}\ln|A'(r)| + \cdots \tag{5.38}$$

where $A(r)$ is given above. Below we concentrate on the solutions with singularity at $r_0 = \infty$, that is (5.35) above.

One can see that the singularity at $r_0 = \infty$ corresponds to an actual curvature singularity in the Einstein frame by computing the Ricci scalar. One finds that R behaves in the Einstein and string frames as

$$R \sim e^{-2A}A'^2, \qquad R_s \sim e^{-2A_s}A_s'^2, \tag{5.39}$$

where R (R_s) are the Einstein (string) frame Ricci scalars. In the latter case A_s is defined in (5.25). For solutions of type (5.35) one finds

$$R \sim e^{2Cr^{\alpha}}r^{2(\alpha-1)}, \qquad R_s \sim \frac{1}{r^{\alpha+1}}. \tag{5.40}$$

Since $\alpha > 1$ we find that there is a curvature singularity in the Einstein frame, but not in the string frame. As we consider these backgrounds as embedded in string theory whose low energy effective action is naturally given in the string frame, we conclude that the IR limit of the holographic theory (5.35) is trustworthy in the sense that the string corrections can indeed be ignored. Instead, R_s diverges near the boundary implying that we can only trust the theory up to a finite UV cut-off r_{UV} determined by demanding $R_s(r_{UV}) \sim \mathcal{O}(1)$. There is another independent curvature invariant with two derivatives, $g^{\mu\nu}\partial_\mu\Phi\partial_\nu\Phi$, and it scales exactly as the Ricci scalar above [5].

5.4   Parameters and the Dilaton Potential

The ihQCD background is further characterized by the choice of parameters in the action and the integration constants that specify the solution to the Einstein's equations. We have a third order system,[9] (6.9). The corresponding three integration

---

[9] These equations are casted in the form of three 1st order equations in Appendix A.

constants can be regarded as the value of the fields Φ, A and X (defined in (5.14)) at a reference point r_f. As mentioned above, see (5.29), the integration constant of X Eq. (5.15) should be fixed by the requirement of an acceptable singularity in the deep interior. As clear from (5.4) $A(r_f)$ scales the volume of the boundary space-time. As we consider QFT with infinite volume we will only be interested in volume independent quantities, i.e. densities, e.g. entropy density, energy density etc. Therefore the integration constant $A(r_f)$ does not enter in physical observables. On the other hand the integration constant $\Phi(r_f)$ is physical and it corresponds to the confinement scale Λ_{QCD} in the dual field theory. This is related to the constant Λ that appears in the UV expansion in (5.21). One can fix this constant of motion either by fitting the actual value of Λ_{QCD} or equivalently by matching the first excited glueball mass, as we discuss below.

In addition, there is the parameter M_p in front of the action (5.1). Action evaluated in a thermal gravitational state corresponds to the free energy of the dual field theory whose free energy in the large T scales as $F \propto T^4$ due to asymptotic freedom and the proportionality constant for pure Yang-Mills fixes M_p as

$$(M_p\ell)^3 = \frac{1}{45\pi^2}, \tag{5.41}$$

as we show in Eq. (2.11). Finally, there is the string length scale ℓ_s that appears in (5.27). It is fixed by matching to the tension of glue flux between the quarks that follows from lattice QCD calculations. This yields

$$\frac{\ell_s}{\ell} \approx 0.15, \tag{5.42}$$

using Eq. (2.11) in the large N limit for $N_f = 0$. This shows that the string α' corrections are suppressed throughout the solution. In particular the massive string states can safely be ignored.[10]

The IR (large exp Φ) and UV (small exp Φ) asymptotics of the dilaton potential is completely fixed by the physical requirements we discussed above. As we also discussed, there is a one-to-one correspondence between the non-perturbative beta function of the field theory and the dilaton potential in the particular renormalization scheme defined by (5.12). In principle, one could be able to fix the entire dilaton potential if one knew the full non-perturbative beta function. Devoid of this information, we take a pragmatic approach and pick a choice which satisfies all the desired features described above:

$$V(\Phi) = \frac{12}{\ell^2} \left\{ 1 + V_0 e^\Phi + V_1 e^{\frac{4}{3}\Phi} \left[\log\left(1 + V_2 e^{\frac{4}{3}\Phi} + V_3 e^{2\Phi} \right) \right]^{1/2} \right\}. \tag{5.43}$$

The 4 parameters of the potential will be determined by comparison to the glueball spectrum and thermodynamic parameters in the next section and the next chapter.

---

[10] Except in the far UV where the holographic model is not applicable.

5.5   The Glueball Spectra

The particle spectrum is obtained by considering finite energy, normalizable excitations around the gravitational background. We first discuss general features of the particle glueball spectra in ihQCD, see [25] for a review of the glueball spectrum calculations in gauge-gravity duality in general.

The action for the fluctuations can be obtained by expanding (5.1) to quadratic order in small perturbations of the background fields[11]

$$ds^2 \rightarrow e^{2A(r)} \left((1+2\varphi)dr^2 + 2A^\mu dr dx^\mu + (\eta_{\mu\nu} + h_{\mu\nu})dx^\mu dx^\nu \right) , \quad (5.44)$$

$$\Phi \rightarrow \Phi + \delta\Phi . \quad (5.45)$$

Five of these degrees of freedom—which we choose to be A_μ and φ—can be gauged away using the diffeomorphisms $x^M \rightarrow x^M + \xi^M$ where $M = 0, 1, 2, 3, r$.

The remaining fields should be put in diffeomorphism invariant combinations. In this review we consider two examples. The first is transverse traceless component of metric fluctuation $\delta g_{\mu\nu}$:

$$\eta^{\mu\nu} h_{\mu\nu}^{TT} = 0, \quad \partial^\mu h_{\mu\nu}^{TT} = 0 . \quad (5.46)$$

These fluctuations determine the spectrum of spin-2 glueballs. A second example concerns spin-0 glueballs, determined by the following diffeomorphism invariant combination [26]

$$\xi = \frac{1}{6}\left(h_\mu^\mu - \frac{\partial^\mu \partial^\nu h_{\mu\nu}}{m^2} \right) - \frac{1}{3X}\delta\Phi , \quad (5.47)$$

where m^2 is the 4D mass of the fluctuation and X is the function defined in (5.14). The quadratic action for fluctuations are obtained by substituting (5.44), (5.45) in (5.1) and expanding up to quadratic order:

$$S[\xi] \sim \int dr d^4 x \, e^{2B(r)} \left[(\partial_r \xi)^2 + (\partial_i \xi)^2 + M^2(r)\xi^2 \right] , \quad (5.48)$$

where $B(r)$ and $M^2(r)$ are functions that are determined by the type of fluctuation which we generically denote by ξ. $M^2 = 0$ for both fluctuations and $B = 3A(r)/2$ and $B = 3A/2 + \log|X|$ for (5.46) and (5.47) respectively. Before specifying to these examples below we discuss features common to all fluctuations

We are interested in 4D mass eigenstates

$$\xi(r, x) = \xi(r)\xi^{(4)}(x), \quad \Box \xi^{(4)}(x) = m^2 \xi^{(4)}(x) , \quad (5.49)$$

---

[11] Alternatively one can fluctuate the background equations of motion to linear order.

for which, on obtains the following fluctuation equation,

$$\ddot{\xi} + 2\dot{B}\dot{\xi} + \Box_4\xi - M^2(r)\xi = 0. \tag{5.50}$$

This equation can be put in Schrodinger form

$$-\frac{d^2}{dr^2}\psi + V_s(r)\psi = m^2\psi, \tag{5.51}$$

with

$$V_s(r) = \frac{d^2B}{dr^2} + \left(\frac{dB}{dr}\right)^2 + M^2(r), \tag{5.52}$$

by redefining

$$\xi(r) = e^{-B(r)}\psi(r). \tag{5.53}$$

We now demand that the energy of the fluctuation ξ is finite. For example the kinetic term in (5.48) yields

$$\left(\int dr|\psi(r)|^2\right)\int d^4x \left(\partial_\mu \xi^{(4)}(x)\right)^2.$$

Demanding that this is finite then leads to the square-integrability condition

$$\int dr|\psi(r)|^2 < \infty. \tag{5.54}$$

in the Schrodinger problem.

Next, we note that the Eq. (5.51) can be written as:

$$\left(P^\dagger P + M^2(r)\right)\psi = m^2\psi, \qquad P = (-\partial_r + \dot{B}(r)). \tag{5.55}$$

This means that the spectrum will be non-negative provided that $M^2 \geq 0$. We now ask the question whether the 4D spectrum is *gapped* or not. If there is a massless mode, $m^2 = 0$ solution to (5.51) then clearly it can only exist for $M^2 = 0$. In this case the solutions to (5.55) can be found analytically

$$\psi_0^{(1)}(r) = e^{B(r)}, \qquad \psi_0^{(2)} = e^{B(r)}\int_0^r dr' e^{-2B(r')}. \tag{5.56}$$

We want to check if these solutions satisfy (5.54). Near the asymptotically AdS boundary we universally have $B \sim 3/2A$ and $A \sim -log(r) + \cdots$. Therefore the first solution above cannot be normalizable near the boundary. The second solution is normalizable near the boundary but not in the deep interior for an arbitrary choice of the dilaton potential [5]. Therefore we cannot find normalizable solutions with $m^2 = 0$.

The only way the mass gap may still vanish is that there is a continuous spectrum starting from $m^2 = 0^+$. This however requires the potential V_s in (5.52) vanishes as $r \to \infty$. From (5.35) we learn that, as $r \to \infty$:

$$A(r) \sim -\left(\frac{r}{R}\right)^{\alpha}, \tag{5.57}$$

therefore

$$V(r) = \dot{B}^2(r) + \ddot{B}(r) \sim R^{-2}\left(\frac{r}{R}\right)^{2(\alpha-1)}. \tag{5.58}$$

We therefore find that *the requirement for gapped spectrum, $\alpha \geq 1$, coincides precisely with the condition for quark confinement*. Note that the two calculations leading to the two conditions are completely independent. Quark confinement comes from analyzing the NG action of the string and mass gap comes from linear fluctuations around the classical background. This is a clear indication that bottom-up holography is capable of demonstrating non-trivial and generic IR properties of confining (large N) gauge theories.

If we require $\alpha > 1$ strictly, we moreover obtain a purely discrete spectrum as the Schrödinger potential (5.52) becomes a well. If $\alpha = 1$ the spectrum becomes continuous for $m^2 \geq V(r \to \infty)$.

Moreover, a WKB analysis of the potential (5.52) [5] gives an asymptotic spectrum for large excitation number $n \gg 1$ as

$$m \sim \Lambda\, n^{\frac{\alpha-1}{\alpha}}. \tag{5.59}$$

In particular we have a linear spectrum for $\alpha = 2$ which is what we choose from now on.

Let us now turn to numerical results. One can determine the glueball spectrum by solving (5.50) numerically. For this one shoots from the boundary starting from the solution in the asymptotically AdS background and demanding that the solution becomes normalizable in the deep interior by tuning m^2. One, hence, finds an infinite set of discrete values for m^2 that corresponds to the glueball spectrum. The result for the few low-lying modes are compared with the lattice results of [27] in the table below. The spin-2 glueball 2^{++} spectrum is obtained from the transverse-traceless fluctuation (5.46) which satisfies (5.50) with $B = 3A/2$ and the spin-0 glueball 0^{++} spectrum with $B = 3A/2 + \log|X|$.

J^{PC}	Lattice (MeV)	Our model (MeV)	Mismatch
0^{++}	*1475 (4%)*	*1475*	0
2^{++}	2150 (5%)	2055	4%
0^{++*}	*2755 (4%)*	*2753*	0
2^{++*}	2880 (5%)	2991	4%
0^{++**}	3370 (4%)	3561	5%
0^{++***}	3990 (5%)	4253	6%

Here the masses in italic are fine-tuned according to the lattice results by fixing the integration constant $A(r_f)$ in Sect. 5.4 and a combination of the parameters V_1 and V_3 in (5.43). The rest are predictions.

References

1. U. Gursoy, E. Kiritsis, JHEP **0802**, 032 (2008). https://doi.org/10.1088/1126-6708/2008/02/032 [arXiv:0707.1324 [hep-th]]
2. K. Skenderis, Class. Quant. Grav. **19**, 5849 (2002). https://doi.org/10.1088/0264-9381/19/22/306 [hep-th/0209067]
3. I. Papadimitriou, JHEP **1108**, 119 (2011). https://doi.org/10.1007/JHEP08(2011)119 [arXiv:1106.4826 [hep-th]]
4. D.Z. Freedman, S.S. Gubser, K. Pilch, N.P. Warner, Adv. Theor. Math. Phys. **3**, 363 (1999). [hep-th/9904017]
5. U. Gursoy, E. Kiritsis, F. Nitti, JHEP **0802**, 019 (2008). https://doi.org/10.1088/1126-6708/2008/02/019 [arXiv:0707.1349 [hep-th]]
6. S.S. Gubser, Adv. Theor. Math. Phys. **4**, 679 (2000). [hep-th/0002160]
7. S.S. Gubser, A. Nellore, Phys. Rev. D **78**, 086007 (2008). https://doi.org/10.1103/PhysRevD.78.086007 [arXiv:0804.0434 [hep-th]]
8. S.S. Gubser, A. Nellore, S.S. Pufu, F.D. Rocha, Phys. Rev. Lett. **101**, 131601 (2008). https://doi.org/10.1103/PhysRevLett.101.131601 [arXiv:0804.1950 [hep-th]]
9. D. Karateev, P. Kravchuk, M. Serone, A. Vichi, JHEP **06**, 088 (2019). https://doi.org/10.1007/JHEP06(2019)088 [arXiv:1902.05969 [hep-th]]
10. N. Nekrasov, S.L. Shatashvili, Phys. Rept. **320**, 127–129 (1999). https://doi.org/10.1016/S0370-1573(99)00059-9 [arXiv:hep-th/9902110 [hep-th]]
11. U. Gürsoy, A. Jansen, W. van der Schee, Phys. Rev. D **94**(6), 061901 (2016). https://doi.org/10.1103/PhysRevD.94.061901 [arXiv:1603.07724 [hep-th]]
12. E. Kiritsis, F. Nitti, L. Silva Pimenta, Fortsch. Phys. **65**(2), 1600120 (2017). https://doi.org/10.1002/prop.201600120 [arXiv:1611.05493 [hep-th]]
13. U. Gürsoy, E. Kiritsis, F. Nitti, L. Silva Pimenta, JHEP **10**, 173 (2018). https://doi.org/10.1007/JHEP10(2018)173 [arXiv:1805.01769 [hep-th]]
14. O. Aharony, S.S. Gubser, J.M. Maldacena, H. Ooguri, Y. Oz, Phys. Rept. **323**, 183 (2000). https://doi.org/10.1016/S0370-1573(99)00083-6 [hep-th/9905111]
15. A.W. Peet, J. Polchinski, Phys. Rev. D **59**, 065011 (1999). https://doi.org/10.1103/PhysRevD.59.065011 [hep-th/9809022]
16. U. Gursoy, E. Kiritsis, L. Mazzanti, F. Nitti, Nucl. Phys. B **820**, 148 (2009). https://doi.org/10.1016/j.nuclphysb.2009.05.017 [arXiv:0903.2859 [hep-th]]
17. U. Gursoy, E. Kiritsis, L. Mazzanti, F. Nitti, JHEP **0905**, 033 (2009). https://doi.org/10.1088/1126-6708/2009/05/033 [arXiv:0812.0792 [hep-th]]
18. E. D'Hoker, D.Z. Freedman, hep-th/0201253
19. J.M. Maldacena, Phys. Rev. Lett. **80**, 4859 (1998). https://doi.org/10.1103/PhysRevLett.80.4859 [hep-th/9803002]

20. S.J. Rey, J.T. Yee, Eur. Phys. J. C **22**, 379 (2001). https://doi.org/10.1007/s100520100799 [hep-th/9803001]
21. Y. Kinar, E. Schreiber, J. Sonnenschein, Nucl. Phys. B **566**, 103 (2000). https://doi.org/10.1016/S0550-3213(99)00652-5 [hep-th/9811192]
22. E. Witten, Adv. Theor. Math. Phys. **2**, 505 (1998) [hep-th/9803131]
23. J. Erlich, E. Katz, D.T. Son, M.A. Stephanov, Phys. Rev. Lett. **95**, 261602 (2005). https://doi.org/10.1103/PhysRevLett.95.261602 [hep-ph/0501128]
24. L. Da Rold, A. Pomarol, Nucl. Phys. B **721**, 79 (2005). https://doi.org/10.1016/j.nuclphysb.2005.05.009 [hep-ph/0501218]
25. R.C. Brower, S.D. Mathur, C.I. Tan, Nucl. Phys. B **587**, 249 (2000). https://doi.org/10.1016/S0550-3213(00)00435-1 [hep-th/0003115]
26. E. Kiritsis, F. Nitti, Nucl. Phys. B **772**, 67–102 (2007). https://doi.org/10.1016/j.nuclphysb.2007.02.024 [arXiv:hep-th/0611344 [hep-th]]
27. H.B. Meyer hep-lat/0508002

Chapter 6
Thermodynamics and the Confinement/Deconfinement Transition

State of the theory at finite temperature is obtained by minimizing the Gibbs free energy F. Gauge-gravity duality relates this to the gravitational action evaluated on-shell [1], that is, by substituting the background in the action (5.1)[1]

$$\frac{F}{T} = S\Big|_{\text{on shell}}. \tag{6.1}$$

This directly follows from Eq. (4.5) by setting $\phi = 0$ (i.e. considering the unperturbed background) and going to Euclidean time. The usual prescription to obtain thermal partition functions in the path integral quantization, i.e. Wick rotating and compactifying time

$$\tau = it, \quad \tau \sim \tau + 1/T, \tag{6.2}$$

and imposing (anti-)periodic boundary conditions on canonical fields, goes through the same in holographic duality. RHS of (6.1) is to be evaluated on a compact time circle. By the assumption of thermal equilibrium, the lagrangian density does not depend on time, hence produce the factor $1/T$ on the LHS.

The next question is, on which background should we compute the action? In general there may be several solutions sharing the same boundary asymptotics. In the theories we consider, generally there exist two such solutions. The first one is obtained by trivially "thermalizing" the ground state solution of the previous chapter, that is, by compactifying the Euclidean time of (5.4). This is called the "thermal gas" solution:

$$ds^2 = e^{2A_0(r)}(dr^2 + d\tau^2 + \delta_{ij}dx^i dx^j), \quad \Phi = \Phi_0(r), \tag{6.3}$$

---

[1] Contribution of the Gibbons-Hawking term to this free energy is on equal footing with the first term in (5.1), it is crucial. Also, without the counterterm the result would generally diverge.

© The Author(s), under exclusive license to Springer Nature Switzerland AG 2021
U. Gürsoy, *Holography and Magnetically Induced Phenomena in QCD*,
SpringerBriefs in Physics,
https://doi.org/10.1007/978-3-030-79599-3_6

with the identification $\tau \sim \tau + 1/T$. The confinement analysis we presented in the previous chapter for the vacuum solution (5.4) goes through for (6.3). Therefore, we can conclude that this solution corresponds to gas of excitations in the confined theory, that is, a gas of glueballs (for pure Yang-Mills) at temperature T. Even though the glueballs are held together by strong interactions the gas itself is weakly coupled because the interactions between glueballs (or hardons in general) are suppressed in the large N limit [2]. The second type of solution is the black-brane solution which we introduce below.

6.1 Black-Brane Solution

The black-brane solution is a black hole with flat horizon:

$$ds^2 = e^{2A(r)} \left(\frac{dr^2}{f(r)} + f(r)d\tau^2 + \delta_{ij}dx^i dx^j \right), \tag{6.4}$$

where theblackening factor $f(r)$ vanishes at a *horizon*:

$$f(r_h) = 0. \tag{6.5}$$

We show below the presence of a horizon is guaranteed (in the class of solutions we consider in this book) once we allow for a non-trivial blackening factor $f \neq 1$ in (6.4). The temperature associated to this black brane solution is determined by Hawking's argument [3] that the solution on the $r - \tau$ plane should involve no conical singularity. This determines the temperature in terms of r_h

$$T = \frac{1}{4\pi} \dot{f}(r_h). \tag{6.6}$$

Near the boundary, the solution (6.4) asymptotes to the AdS-Schwarzschild black brane,

$$A(r) \rightarrow -\log(r/\ell), \qquad f(r) \rightarrow 1 - \frac{r^4}{r_h^4}, \qquad \text{as } r \rightarrow 0, \tag{6.7}$$

for which, the relation between temperature and r_h is given by

$$T = 1/\pi r_h \qquad \text{as } r_h \rightarrow 0, \tag{6.8}$$

What state in QFT does the black-brane solution correspond to? Following the discussion on confinement in Sect. 5.3 in Chap. 5 we learn that this solution corresponds to a state with *deconfined* color charges. This is because as you pull the end points of the test quarks apart from each other the tip of the string will move towards the interior of the geometry and above a certain separation length $L = L_{max}$ the tip of the string will touch the horizon and dissolve. Therefore the potential energy between test quarks will not be linear in L for sufficiently large L.

We show below that, the black brane solution dominates over thermal gas in the thermal ensemble for sufficiently high temperatures. This is a general result from holography [4]. Furthermore, high T in this regime corresponds to small values of r_h. This allows us to prove (assuming the holographic description of QFT is valid) the following generic non-trivial result: *if the gauge theory confines color charge in the ground state, then there exists a confinement/deconfinement transition at a finite temperature.* Even though plausible, a direct QFT proof of this statement is quite difficult [5]. It can be proven in compact U(1) gauge theory using analytic techniques [6].

In the large-N limit we consider here, in which the flavor degrees of freedom are suppressed, the black brane is dual to a plasma of gluons. In Chap. 7 we consider another, the so-called "Veneziano limit", where the number of flavors are also taken to infinity. In that case, black-brane will correspond to a plasma of gluons and quarks.

Einstein's equations evaluated on the ansatz (6.4) read,

$$\ddot{A} - (\dot{A})^2 = -\frac{4}{9}(\dot{\Phi})^2, \quad 3\ddot{A} + 9(\dot{A})^2 + 3\dot{A}\frac{\dot{f}}{f} = \frac{e^{2A}}{f}V(\Phi), \quad \ddot{f} + 3\dot{A}\dot{f} = 0.$$
(6.9)

We note that these equations reduce to (5.5) when one sets $f = 1$, as in (6.3). Let us now count the number of constants of motion in this problem, which parametrizes these black-brane solutions. This can be done most efficiently by reformulating Einstein's equations in terms of scalar variables, just as in (5.14). There are two such independent scalar variables:

$$X(\Phi) = \frac{1}{3}\frac{\dot{\Phi}(r)}{\dot{A}(r)}, \quad Y = \frac{1}{4}\frac{\dot{f}(r)}{f(r)\dot{A}(r)}.$$
(6.10)

As shown in Appendix A, the Einstein's equations then reduce to a coupled first order system[2]:

$$\frac{dX}{d\Phi} = -\frac{4}{3}(1 - X^2 + Y)\left(1 + \frac{3}{8}\frac{1}{X}\frac{d\log V}{d\Phi}\right),$$
(6.11)

$$\frac{dY}{d\Phi} = -\frac{4}{3}(1 - X^2 + Y)\frac{Y}{X}.$$
(6.12)

Assuming a regular horizon at r_h where f vanishes as a power of $r - r_h$, it is clear from the definition in (6.10) Y should diverge at the horizon. This is because f vanishes while $\dot{A}$ and $\dot{f}$ should stay finite.[3] Φ should also be finite at the horizon

---

[2] As shown in Appendix A once X and Y are determined the background functions Φ, A and f can be obtained by a single integration.

[3] If $\dot{A}$ diverged this would give rise to a curvature singularity at the horizon. $\dot{f}$ cannot vanish if T is finite by (6.6).

which means that Y should diverge like $(r_h - r)^{-1} \sim (\Phi_h - \Phi)^{-1}$ where Φ_h is the value of dilaton at the horizon. Finally, $dX/d\Phi$ should also stay finite at the horizon to avoid a curvature singularity there. We then find from (6.11) that value of X at the horizon is completely fixed as

$$X_h \equiv X(\Phi_h) = -\frac{3}{8} \frac{d \log V}{d\Phi}\bigg|_{\Phi_h} \tag{6.13}$$

which then, using (6.12) also determines the behavior of Y near the horizon as,

$$Y \to \frac{Y_h}{\Phi_h - \Phi}, \qquad Y_h = -\frac{3}{4} X_h . \tag{6.14}$$

This means that, regularity at the horizon fixes one of the integration constants in the system (6.11), (6.12), leaving a single integration constant that is Φ_h. This constant is related to the temperature of the system as we show below. Solving the rest of the first order differential equations for Φ, A and f as in Appendix A we find 3 more integration constants. For asymptotically AdS space we need to require $f \to 1$ at the boundary, see (6.7) which fixes one of these constants. The integration constant for the Φ equation can be identified with λ_{QCD} in the dual theory, just as in the previous chapter. Finally, the integration constant of the A equation is again related to the volume of the dual theory which is scaled away in dimensionless quantities. Hence we obtain only two non-trivial integration constants, Λ_{QCD} and T. Λ_{QCD} of the black-brane solution should be identified with that of the ground state solution (or thermal gas) these correspond to two different states in the same theory. We conclude that thermodynamic functions, such as the free energy, only depend on the dimensionless combination T/Λ_{QCD}.

We can now expand on the *good singularity* condition mentioned below Eq. (5.30). We claimed there that this condition uniquely fixes the IR asymptotics of the confined solution to be (5.30). As demonstrated in [7] a strong version of the good singularity condition requires that the TG solution be obtained from a BB solution in the limit the horizon marginally traps the singularity. From (6.13) we observe that the value of X at the horizon is completely fixed in terms of the dilaton potential. Hence sending $r_h \to \infty$ ($\Phi_h \to \infty$) we discover (5.30) as the correct condition to be satisfied by the TG solution, to have a good singularity, that can be marginally trapped by an infinitesimal horizon.

Finally, we present an argument as to why the black brane ansatz (6.4) requires presence of a horizon. From (6.12) it is clear that $dY/d\Phi \geq 0$. As $Y > 0$ near the boundary, it will stay positive throughout the solution. Using (6.10) this requires $\dot{f} < 0$ using $\dot{A} < 0$ which we established before, and the condition that $f \to 1$ near the boundary. Using (6.9) this means $\ddot{f} < 0$ as well. Assuming smoothness, such a function necessarily goes through a zero. If this zero r_h is not at a finite point then it should be at $r_h \to \infty$, equivalently $\Phi_h = \infty$. Taking this limit in (6.13) and (6.14), we learn that the black brane exactly reduces to the thermal gas solution.[4] Thus, if the black brane solution is assumed to be distinct from the thermal gas, its blackening function should vanish at finite r_h.

6.2 Entropy, Gluon Condensate and Conformal Anomaly

The entropy of the black brane is given by the Bekenstein-Hawking [3, 8] formula as the area of the horizon:

$$S = \frac{\text{area}}{4G_N} = 4\pi M_p^3 N_c^2 e^{3A(r_h)} V_3 , \tag{6.15}$$

where V_3 is the spatial volume spanned by coordinates x, y, z and we used the relation $16\pi G_N = 1/(M_p^3 N_c^2)$. At large T, the BB solution approaches to AdS with $\exp A \to \ell/r$ resulting in

$$S/T^3 \to 4\pi^4 (M_p \ell)^3 N_c^2 V_3, \qquad T \to \infty , \tag{6.16}$$

where we used (6.8).

Entropy, as well as other thermodynamic quantities, can also be read off from the subleading terms near the boundary.

$$e^{A(r)} = e^{A_0(r)} \left(1 + G\frac{r^4}{\ell^3} + \cdots \right), \qquad r \to 0 \tag{6.17}$$

$$\Phi(r) = \Phi_0(r) + \frac{45G}{8}\frac{r^4}{\ell^3} \log \Lambda r + \cdots , \qquad r \to 0 \tag{6.18}$$

$$f(r) = 1 - \frac{C}{4}\frac{r^4}{\ell^3} + \cdots , \qquad r \to 0 \tag{6.19}$$

Here G and C are integration constants of the black-brane that depend on r_h. Solving the last equation in (6.9) we find

$$f(r) = 1 - \frac{\int_0^r e^{-3A(r)} dr}{\int_0^{r_h} e^{-3A(r)} dr} , \tag{6.20}$$

where we fixed the integration constants in the solution requiring $f \to 1$ at the boundary and requiring that it vanishes at r_h. Expanding this near the boundary, where $\exp A \to 1/r$, and comparing to (6.19) we find

---

[4] More precisely f becomes a step function, which is 1 everywhere except at $r_h = \infty$ where it drops to zero.

$$C = \frac{1}{\int_0^{r_h} e^{-3A(r)} dr} .$$

On the other hand, using the formulae for the temperature and entropy in (6.6) and (6.15) we obtain

$$\frac{1}{\int_0^{r_h} e^{-3A(r)} dr} = TS/(M_p^3 N_c^2 V_3) .$$

Thus, we find that C in (6.19) is the *enthalpy density*

$$C = Ts/M_p^3 , \tag{6.21}$$

where we define s as the entropy density per gluon $s = S/N_c^2 V_3$.

Working out the meaning of G in (6.17) and (6.18) is also straightforward. As it arises from the difference of the normalizable terms in Φ and Φ is dual to the operator $\operatorname{tr} F^2$, it is identified with the difference of the VeVs of this operator between the plasma state and the confined state. A careful calculation (see Sect. 4 of [4]) yields:

$$\langle \operatorname{tr} F^2 \rangle_{BB} - \langle \operatorname{tr} F^2 \rangle_{TG} = -\frac{240}{b_0} M_p^3 N_c^2 \, G , \tag{6.22}$$

where b_0 enters through (5.21). Alternatively, one can interpret G as the excess of conformal anomaly between the plasma and confined phases [4]:

$$\langle T_\mu^\mu \rangle_{BB} - \langle T_\mu^\mu \rangle_{TG} = 60 M_p^3 N_c^2 \, G . \tag{6.23}$$

One can check that the expressions (6.22) and (6.23) obeys the expected Ward identity[5]

$$T_\mu^\mu = \frac{\beta(\lambda)}{4\lambda^2} \operatorname{tr} F^2 \tag{6.24}$$

near UV [4]. We learn that G is the gluon condensate in the plasma phase normalized by its vacuum value.

6.3   Deconfinement Transition

The thermal state of the theory at a given T is given by whichever of (6.3) or (6.4) minimizes the free-energy. This can be inferred from the difference of on-shell actions

$$\Delta S = S[\text{sol2}] - S[\text{sol1}] = \Delta F/T . \tag{6.25}$$

---

[5] This is a non-trivial check of the overall holographic construction.

When $\Delta S < 0$ $(\Delta S > 0)$ the plasma (confined) state wins. Calculation of this difference is non-trivial. Here I will only highlight important points sparing the reader from details which can be found in Appendix C of [4]. First, one has to note that there are two contributions to the difference, the one coming from the Einstein-Hilbert action and the other from the Gibbons-Hawking term in (5.1) (the counter-term piece in (5.1) precisely cancels in the difference). Thus, we write each term as $S = S_{EH} + S_{GH}$. Both contributions can be expressed in terms of the boundary asymptotics of the background functions. This is obvious for the GH term (5.2), but also true for the EH term. The latter is because, using equations of motion, one can express the integrand in the EH term as a total derivative:

$$S_{EH} = 2M_p^2 V_3 \beta \int_\epsilon^{r_h} \frac{d}{dr} \left(\dot{A}(r) f(r) e^{3A(r)} \right) . \qquad (6.26)$$

where the $\beta = 1/T$ factor arises from the Euclidean time integral and V_3 the volume of boundary space arising from the spatial integration. ϵ is a UV cut-off which we will remove eventually. The contribution from the horizon vanishes as $f(r_h) = 0$ and the other background functions are finite. Thus,

$$S_{EH} = -2M_p^2 V_3 \beta \dot{A}(\epsilon) f(\epsilon) e^{3A(\epsilon)}. \qquad (6.27)$$

On the other hand the GH term follows from substituting in (5.2) the metric ansatz:

$$S_{GH} = M_p^2 V_3 \beta e^{3A(\epsilon)} f(\epsilon) \left(8\dot{A}(\epsilon) + \frac{\dot{f}(\epsilon)}{f(\epsilon)} \right) . \qquad (6.28)$$

Both (6.28) and (6.27) are divergent in the limit $\epsilon \to 0$. This in the dual QFT corresponds to the UV divergence of the bubble diagrams contributing to free energy. This divergence is cancelled in the difference (6.25) because both states should display the same UV asymptotics. The gravity dual of this statement is that one must demand background functions becoming the same near the boundary. Comparison of the metrics (6.3) and (6.4) then yields

$$\beta_0 e^{A_0(\epsilon_0)} = \beta e^{A(\epsilon)} \sqrt{f(\epsilon)}, \qquad V_3^0 e^{3A_0(\epsilon_0)} = V_3 e^{3A(\epsilon)}, \qquad \Phi_0(\epsilon_0) = \Phi(\epsilon) , \quad (6.29)$$

which arise from matching the time-cycles, the space-cycles and the dilatons at the cut-offs and we allowed for different values for length of these cycles and position of the cut-offs in the black-brane and the thermal gas solutions.[6] One can now calculate the difference (6.25) using (6.27) and (6.28) both for the BB and the TG solutions, requiring (6.29), substituting the near boundary expansions (6.17), (6.18) and (6.19) and taking the limit $\epsilon \to 0$ to obtain a finite result:

---

[6] The latter is necessary in order to leave freedom to keep the integration constants Λ in (5.21) and the analogous UV expansion of the black-brane function the same [4].

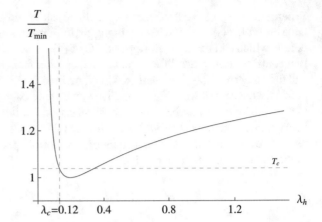

Fig. 6.1 Temperature of the black brane solution as a function of $\exp \Phi_h$ in the ihQCD model

$$\Delta F = \frac{1}{\beta} \Delta S = M_p^3 N_c^2 V_3 \left(15 G(T) - \frac{1}{4} T s \right) . \tag{6.30}$$

One can also calculate the energy difference in the two states using the ADM mass formula (see [4] for details) as

$$\Delta E = M_p^3 N_c^2 V_3 \left(15 G(T) + \frac{3}{4} T s \right) . \tag{6.31}$$

Combining (6.30) and (6.31) we learn that the system satisfies the Smarr relation $F = E - TS$ as it should. We finally note that the functions G and s in the expression for the free energy depends on the integration constant r_h (or Φ_h) as they are obtained from the near boundary expansions of the background functions. To express this result in T one still has to relate r_h (or Φ_h) to T. This can be done by calculating (6.6). The result is show in Fig. 6.1 where we show T as a function of $\exp \Phi_h$ for convenience.

Two comments are in order. First, we see that the black-brane exists above a minimum temperature $T = T_{min}$ that depends on the particular model. Below this temperature there exists only the thermal gas solution and it dominates the ensemble. Second, we see that for any $T > T_{min}$ there are actually two black brane branches one with a large value of ϕ_h (or r_h) and one with a small value of ϕ_h. The black-brane with smaller value of r_h has a bigger event horizon since, as we showed in the previous section, $A(r_h)$ is a monotonically decreasing function and the event horizon is proportional to $\exp 3A(r_h)$. Therefore, we call the solution with smaller r_h the *large black-brane* and the solution with larger r_h the *small black-brane*. As we show below, the latter solution is always subdominant in the ensemble, whereas the former one, the large black-brane corresponds to the true plasma phase in the theory.

We can finally answer our original question: which phase minimizes F at a given T? In Eq. (6.30), the gluon condensate G is a positive definite quantity. On the other hand the entropy term is negative definite. It is therefore conceivable that there exists a critical temperature T_c where ΔF vanishes. At very high temperatures the

Fig. 6.2 Difference of free energies between the plasma and the confined phase ΔF as a function of T

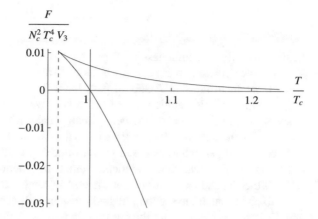

entropy term normalized by T^4 should go to a positive constant given by (6.16). On the other hand the difference in the gluon condensate $G(T)/T^4$ should vanish as it should approach the same value in the plasma and the confined phases in the UV. This means that at large T the plasma phase wins. The question then is, whether or not ΔF becomes positive at small T. The answer is in the affirmative and can be obtained by calculating (6.30) numerically as in [9]. One then obtains Fig. 6.2 for the free energy. The x-axis in this figure, $F/N^2 = 0$, corresponds to the free energy of the thermal gas solution. This is because, as discussed above, the thermal gas solution is obtained by sending $r_h \to \infty$ (on the small BB branch) in Fig. 6.1. In this limit the horizon area shrinks to zero yielding vanishing entropy. Similarly the ADM mass of the black-brane also vanishes yielding vanishing E. Then from the Smarr formula we have $F(TG)/N^2 \to 0$. We notice the presence of two black-brane branches in this figure. They exist above $T = T_{min}$ and the one with positive F is the small black-brane. As clear from the figure this branch is always sub-dominant in the ensemble. The lower branch is the large black-brane branch. It crosses the x-axis at $T = T_c$ which is higher than T_{min}. Therefore we obtain the holographic description of the *deconfinement transition* in our holographic model. We also observe that this is a *first order* phase transition as expected in large N QCD.

An insightful use of holography is to prove general statements pertaining the dual QFT. As advertised above, presence of the confinement/deconfinement transition is such an example. This follows from the following geometric argument [4]. First we show that T as a function of r_h always possesses a (single) minimum, see Fig. 6.1, in a theory with a confining ground state. This follows from the fact that T diverges both as $r_h \to 0$ and as $r_h \to \infty$. The former comes from the requirement of asymptotic AdS, the latter comes from use of (6.6) and (6.20) together with the IR asymptotics (5.35) with $\alpha > 1$ required for confinement.[7] Thus T has a minimum at r_{min} if the theory confines in the ground state. On the other hand, the first law determines entropy as

---

[7] This does not guarantee a single minimum but instead an odd number of extrema, which is also sufficient to prove what follows. We will assume a single minimum for simplicity.

$$S = -\frac{dF}{dT} = -\frac{dF/dr_h}{dT/dr_h} .$$

From positivity and finiteness of S on the BB branch(s) we discover that $F(r_h)$ should have a maximum at r_{min}. As F should vanish as $r_h \to \infty$ and $F \to -\infty$ as $r_h \to 0$ we then find that F should pass from 0 at a locus $r_c < r_{min}$. This proves that a confining holographic theory (of the type we describe here) is guaranteed to possess a confinement/deconfinement transition at finite T_c with $\infty > T_c > T_{min}$.

To obtain the free energy in (6.2) we fixed the parameters of the model as explained in Sect. 5.4. In particular, as mentioned around Eq. (5.20) V_0 and $V_2 V_1$ is fixed by matching the first two scheme-independent beta-function coefficients in the pure $SU(N)$ theory, and as mentioned at the end of Sect. 5.5, a combination of V_1 and V_3 is fixed to match the second glueball mass in [10]. The other combination of V_1 and V_3 is fixed by matching the entropy density with the lattice result of [11] at a fixed temperature $T = 2T_c$. The best fit turns out to be $V_1 = 14$, $V_3 = 170$. The only non-trivial integration constant (apart from T) in the background solutions is Λ that is fixed by matching the first glueball mass as explained in Sect. 5.5. Having fixed

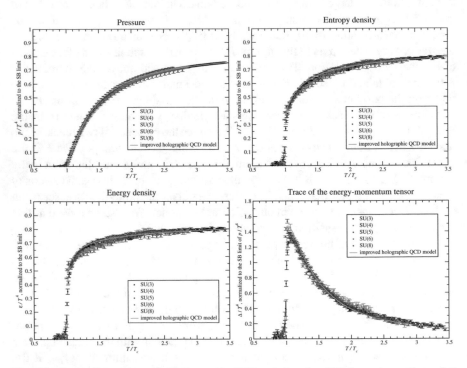

Fig. 6.3 Comparison of improved holographic QCD to lattice data [13] for various values of N with ihQCD (orange curves). All thermodynamic functions are densities normalized by a factor of N^2 and further by an appropriate power of T to make them dimensionless. Pressure $P = -F$ for extensive systems. Reprinted with permission from APS publishers

all the parameters in the model, the rest are predictions to be tested against lattice data. In particular one obtains

$$T_c = 247\,\text{MeV} \tag{6.32}$$

for the transition temperature, that compares very well with the lattice data [12]. For the latent heat $L_h = \Delta E(T_c) = T_c \Delta S(T_c)$ at the transition we find

$$L_h = 0.31 N_c^2 T_c^4\,, \tag{6.33}$$

that also agrees with the lattice data at large N_c [12]. Reference [13] studied the thermodynamic functions of pure $SU(N)$ theory at various values of N_c and compared his data with our findings, see Fig. 6.3.

We observe two important features in these plots. First, when appropriately normalized, the thermodynamic quantities collapse on a single curve, almost independent of N. This implies that holography which necessarily requires $N \to \infty$ is not necessarily far from $N = 3$, at least, as far as thermodynamics is concerned. Second, we observe that the thermodynamic functions coming from the ihQCD model matches astonishingly well.

References

1. E. Witten, Adv. Theor. Math. Phys. **2**, 505 (1998) [hep-th/9803131]
2. S. Coleman, Aspects of symmetry: selected erice lectures. https://doi.org/10.1017/CBO9780511565045
3. S.W. Hawking, Phys. Rev. D **13**, 191–197 (1976). https://doi.org/10.1103/PhysRevD.13.191
4. U. Gursoy, E. Kiritsis, L. Mazzanti, F. Nitti, JHEP **0905**, 033 (2009). https://doi.org/10.1088/1126-6708/2009/05/033 [arXiv:0812.0792 [hep-th]]
5. D.J. Gross, R.D. Pisarski, L.G. Yaffe, Rev. Mod. Phys. **53**, 43 (1981). https://doi.org/10.1103/RevModPhys.53.43
6. A.M. Polyakov, Phys. Lett. B **72**, 477–480 (1978). https://doi.org/10.1016/0370-2693(78)90737-2
7. S.S. Gubser, Adv. Theor. Math. Phys. **4**, 679 (2000). [hep-th/0002160]
8. J.D. Bekenstein, Phys. Rev. D **7**, 2333–2346 (1973). https://doi.org/10.1103/PhysRevD.7.2333
9. U. Gursoy, E. Kiritsis, L. Mazzanti, F. Nitti, Nucl. Phys. B **820**, 148 (2009). https://doi.org/10.1016/j.nuclphysb.2009.05.017 [arXiv:0903.2859 [hep-th]]
10. H.B. Meyer. hep-lat/0508002
11. G. Boyd, J. Engels, F. Karsch, E. Laermann, C. Legeland, M. Lutgemeier, B. Petersson, Nucl. Phys. B **469**, 419 (1996). https://doi.org/10.1016/0550-3213(96)00170-8 [hep-lat/9602007]
12. B. Lucini, M. Teper, U. Wenger, JHEP **0502**, 033 (2005). https://doi.org/10.1088/1126-6708/2005/02/033 [hep-lat/0502003]
13. M. Panero, Phys. Rev. Lett. **103**, 232001 (2009). https://doi.org/10.1103/PhysRevLett.103.232001 [arXiv:0907.3719 [hep-lat]]

Chapter 7
Improved Holographic QCD at Finite Density

So far we considered the holographic theory only in the glue sector. Our description with the limit $N \to \infty$ applies both to pure Yang–Mills theory, where the only microscopic excitations are gluons, and to large-N QCD with *finite number of flavors* N_f. The latter case is included because one can consistently ignore fermion loop corrections to Feynman diagrams when the ratio N_f/N is parametrically suppressed. In real QCD, however, contribution of quarks are as important as gluons. At energy scales of order Λ_{QCD} one typically considers $N_f = 2$ or 3 light flavors: up, down and strange quarks, and, number of colors $N = 3$ with their ratio $N_f/N = 2/3$ or 1. To mimic this balance between quarks and gluons in the large N limit, one typically considers the *Veneziano limit* [1] by also taking also the number of flavors to infinity, keeping their ratio finite:

$$N_f, N \to \infty, \quad x = \frac{N_f}{N} = \text{fixed}, \quad \lambda = \frac{g_{\text{YM}}^2 N}{8\pi^2} = \text{fixed}. \qquad (7.1)$$

We keep the ratio x as a free parameter in what follows, the actual value for real QCD with light flavors corresponding to $x = 1$ (for up, down and strange) or $x = 2/3$ (for up and down quarks). As discussed in Chap. 2, an important operator characterizing the quark sector is the quark condensate and our discussion will focus on how to implement it in holography and how its presence affects the theory in the strong coupling limit.

In passing, we note that QCD exhibits a very rich IR structure as a function of the ratio x, including presence of a conformal window in a range of x between $x_c \sim 4$ and $11/2$, a Berezinskii–Kosterlitz–Thouless type phase transition [2] between chiral symmetry breaking and conformal phases at x_c [3], and a weakly coupled IR fixed point at $x = 11/2$ [4]. This ground state structure in the Veneziano limit is explored in detail using the holographic model [5] which we present below. Here, we focus on the finite temperature aspects of this holographic theory.

© The Author(s), under exclusive license to Springer Nature Switzerland AG 2021
U. Gürsoy, *Holography and Magnetically Induced Phenomena in QCD*,
SpringerBriefs in Physics,
https://doi.org/10.1007/978-3-030-79599-3_7

Improved holographic QCD has been extended in [5] to include quarks in the Veneziano limit. Quarks are introduced through space-filling 5D flavor branes [6–8]. These are space-filling N_f D4-branes and N_f $\bar{D}$4-branes embedded in the 5D bulk geometry. In the Veneziano-limit, the energy-momentum tensor of these flavor branes become comparable to the Planck mass $M_p^3 N^2$ in (5.1), hence they back-react and deform the background. Therefore one must solve Einstein's equations obtained from the full action:

$$S = S_g + S_f, \tag{7.2}$$

where the glue part S_g, is given in (5.1) and the flavor part, S_f, consists of the DBI action of $N_f + \bar{N}_f$ space-filling D4 brane-anti-brane pairs for N_f left and right handed quarks and their Wess–Zumino coupling to the background Ramond–Ramond fields. This Wess–Zumino action includes a Chern–Simons term which plays important role for realization of QCD anomalies in holography [8]. We will ignore the Wess–Zumino contribution in our presentation for simplicity, assuming its effects are implemented. See [9, 10] for a detailed discussion of its relevance to anomalies and transport in QCD. Our main focus in the flavor section is the DBI action of flavor branes, which reads [5, 8]:

$$S_f = -\frac{1}{2} M_p^3 N_c \mathbb{Tr} \int d^5 x \left(V_f(\lambda, T^\dagger T) \sqrt{-\det \mathbf{A}_L} + V_f((\lambda, TT^\dagger) \sqrt{-\det \mathbf{A}_R} \right), \tag{7.3}$$

where $\mathbb{Tr}$ denotes "symmetric-trace" on non-Abelian branes, defined in [11] as $\mathbb{Tr}(a_1 \cdots a_n) = \frac{1}{n!} \mathrm{Tr}(a_1 \cdots a_n + \text{permutations})$. The fields $\mathbf{A}$ are given by

$$\mathbf{A}_{L\mu\nu} = g_{\mu\nu} + w(\lambda, T) F_{\mu\nu}^L + \frac{\kappa(\lambda, T)}{2} \left[(D_\mu T)^\dagger (D_\nu T) + (D_\nu T)^\dagger (D_\mu T) \right],$$

$$\mathbf{A}_{R\mu\nu} = g_{\mu\nu} + w(\lambda, T) F_{\mu\nu}^R + \frac{\kappa(\lambda, T)}{2} \left[(D_\mu T)(D_\nu T)^\dagger + (D_\nu T)(D_\mu T)^\dagger \right], \tag{7.4}$$

with the covariant derivatives defined as

$$D_\mu T = \partial_\mu T + i T A_\mu^L - i A_\mu^R T. \tag{7.5}$$

Here, A_L and A_R denote gauge fields living on the flavor D-branes corresponding to the global flavor symmetry $U(N_f)_L \times U(N_f)_R$ with F^L and F^R corresponding field strengths. T is a complex scalar, called the *open string tachyon*, that transforms in bifundamental representation of the flavor symmetry and holographically corresponds to the quark mass operator $\bar{q}q$. Following[1] [6–8] we choose the tachyon potential as

$$V_f(\lambda, TT^\dagger) = V_{f0}(\lambda) e^{-a(\lambda) TT^\dagger}. \tag{7.6}$$

This form of the tachyon action was motivated in [6–8] by reproducing the expected spontaneous symmetry breaking and the axial anomaly of QCD.

---

[1] An inspiration behind these actions is Sen's work on the open string tachyon [12].

The theory, in its final form, has been put forward in [5] where it was coined "V-QCD" in [5] for "V" standing for "Veneziano", and further developed in the subsequent works [9, 13–18]. Then V_{f0}, w and κ are new potentials (in addition to V in (5.1)) that, in the bottom-up approximation should be determined by phenomenological requirements explained in Chap. 2. A judicious choice for these potentials are presented in Appendix A.2.

In V-QCD, one typically makes a simplifying assumption by taking $\kappa(\lambda, T)$ and $w(\lambda, T)$ independent of T. For equal quark masses (that we take zero in this section) for all N_f flavors, one can further simplify the action by choosing a diagonal tachyon field

$$T = \tau(r)\mathbb{I}_{N_f}, \tag{7.7}$$

that corresponds to N_f light quarks with the same mass in boundary field theory. As mentioned above, $\tau(r)$ is holographically dual to the quark mass operator and its non-trivial profile is responsible for the chiral symmetry breaking on the boundary theory. The boundary asymptotics of this function, for the choice of potentials given in Appendix A.2 is

$$\tau(r) \simeq m_q r(-\log \Lambda r)^{-\rho} + \langle \bar{q}q \rangle r^3 (-\log \Lambda r)^\rho \tag{7.8}$$

the power ρ is to be matched to the anomalous dimension of $\bar{q}q$ and the QCD β-function (see [5, 14] for details). In this review we only consider massless quarks $m_q = 0$ so the non normalizable mode of the tachyon solution vanishes, thus providing a boundary condition for the τ equation of motion.

Flavor current correlators in the holographic theory follow from fluctuating the bulk gauge fields A_L^a and A_R^a in (7.3) where the small a index corresponds to non-Abelian flavor. We will not discuss these here further. Instead we focus on the effects of non-vanishing quark chemical potential μ on the phase diagram. This chemical potential can be introduced through the boundary value of the time component of $a = 0$, part of the $U(1)_V$ bulk gauge fields

$$A_v^V = \frac{A_{L,v}^0 + A_{R,v}^0}{2} \to (\mu + qr^2 + \cdots, 0, 0, 0, 0), \quad r \to 0, \tag{7.9}$$

where q is proportional to the charge conjugate to this chemical potential. Accordingly, we will further simplify the flavor action by setting all A_L and A_R to zero except (7.9):

$$S_f = -x M^3 N_c^2 \int d^5x \, V_f(\lambda, \tau) \sqrt{-\det \left(g_{\mu\nu} + w(\lambda) F_{\mu\nu}^V + \kappa(\lambda) \partial_\mu \tau \partial_\nu \tau \right)}, \tag{7.10}$$

where $\lambda = \exp(\Phi)$ is the dilaton.

In what follows, we focus on the qualitative picture arising from (7.10) and (5.1), in particular the influence of the quark chemical potential on the phase diagram Fig. 7.1 [15]. We observe the possibility of three phases in this diagram. The confined

Fig. 7.1 The phase diagram
of the ihQCD theory in the
Veneziano limit with finite
quark chemical potential and
vanishing quark masses.
Axes labels are given in units
of Λ_{QCD}. Figure taken from
[15]

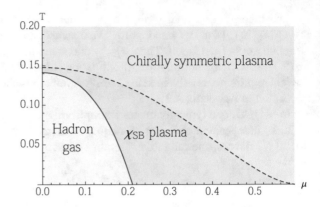

phase, denoted by "hadron gas" in the figure, continues to exist up to some finite
$\mu \neq 0$ in the small temperature regime. Chiral symmetry is spontaneously broken in
this phase as $SU(N_f)_L \times SU(N_f)_R \rightarrow SU(N_f)_{L+R}$ due to the finite VeV of quark
condensate. This phase holographically corresponds to the thermal gas solution in
the previous section, generalized to $\mu \neq 0$. We observe two other phases for larger
values of the temperature and μ. The phase denoted by χ_{SB} is a deconfined quark-
gluon plasma with a non-vanishing value of the quark condensate, therefore the
chiral symmetry remains broken there. Holographically, this phase corresponds to
the black-brane phase of the previous section accompanied by a non-trivial vector
bulk field (7.9) and a non-trivial profile for the tachyon field $\tau(r)$. The hadron gas
phase is separated from the χSB phase by a first order phase separation curve $T_c(\mu)$
(red, solid) in Fig. 7.1. Heating the system up, the quark-condensate melts through
a second-order phase transition (blue, dashed curve) at $T_\chi(\mu)$ and one obtains a
deconfined state where the chiral symmetry is restored. The phase boundary becomes
a continuous crossover for finite quark masses [15]. The high temperature (pink
in Fig.) phase holographically corresponds to a generalization of the black-brane
background of the previous section for finite (7.9) and the open string tachyon $\tau = 0$
in this phase. These generic qualitative features of the phase diagram remains the
same for different choices of the potentials in the action, in particular the existence of
the three phases and the nature of the phase boundaries remain unaltered. However,
the precise location of the phase boundaries will depend on the details of the bottom-
up construction. We add a third axis to this phase diagram, an external magnetic field,
in the next chapter.

References

1. G. Veneziano, Nucl. Phys. B **159**, 213–224 (1979). https://doi.org/10.1016/0550-3213(79)90332-8
2. J.M. Kosterlitz, D.J. Thouless, J. Phys. C **6**, 1181–1203 (1973). https://doi.org/10.1088/0022-3719/6/7/010

3. V.A. Miransky, Nuovo Cim. A **90**, 149–170 (1985). https://doi.org/10.1007/BF02724229
4. T. Banks, A. Zaks, Nucl. Phys. B **196**, 189–204 (1982). https://doi.org/10.1016/0550-3213(82)90035-9
5. M. Jarvinen, E. Kiritsis, JHEP **03**, 002 (2012). https://doi.org/10.1007/JHEP03(2012)002, arXiv:1112.1261 [hep-ph]
6. F. Bigazzi, R. Casero, A.L. Cotrone, E. Kiritsis, A. Paredes, JHEP **0510**, 012 (2005). https://doi.org/10.1088/1126-6708/2005/10/012, arXiv:hep-th/0505140
7. R. Casero, C. Nunez, A. Paredes, Phys. Rev. D **73**, 086005 (2006). https://doi.org/10.1103/PhysRevD.73.086005, arXiv:hep-th/0602027
8. R. Casero, E. Kiritsis, A. Paredes, Nucl. Phys. B **787**, 98 (2007). https://doi.org/10.1016/j.nuclphysb.2007.07.009, arXiv:hep-th/0702155 [HEP-TH]
9. D. Arean, I. Iatrakis, M. Jarvinen, E. Kiritsis, arXiv:1609.08922 [hep-ph]
10. U. Gürsoy, A. Jansen, JHEP **10**, 092 (2014). https://doi.org/10.1007/JHEP10(2014)092, arXiv:1407.3282 [hep-th]
11. A.A. Tseytlin, Nucl. Phys. B **501**, 41–52 (1997). https://doi.org/10.1016/S0550-3213(97)00354-4, arXiv:hep-th/9701125 [hep-th]
12. A. Sen, Int. J. Mod. Phys. A **20**, 5513 (2005). https://doi.org/10.1142/S0217751X0502519X, arXiv:hep-th/0410103
13. D. Arean, I. Iatrakis, M. Järvinen, E. Kiritsis, Phys. Lett. B **720**, 219 (2013). https://doi.org/10.1016/j.physletb.2013.01.070, arXiv:1211.6125 [hep-ph]
14. D. Areán, I. Iatrakis, M. Järvinen, E. Kiritsis, JHEP **1311**, 068 (2013). https://doi.org/10.1007/JHEP11(2013)068
15. T. Alho, M. Järvinen, K. Kajantie, E. Kiritsis, C. Rosen, K. Tuominen, JHEP **1404**, 124 (2014). Erratum: [JHEP **1502**, 033 (2015)]. https://doi.org/10.1007/JHEP02(2015)033, https://doi.org/10.1007/JHEP04(2014)124, arXiv:1312.5199 [hep-ph]
16. I. Iatrakis, I. Zahed, JHEP **1504**, 080 (2015). https://doi.org/10.1007/JHEP04(2015)080, arXiv:1410.8540 [hep-th]
17. T. Alho, M. Jarvinen, K. Kajantie, E. Kiritsis, K. Tuominen, Phys. Rev. D **91**(5), 055017 (2015). https://doi.org/10.1103/PhysRevD.91.055017, arXiv:1501.06379 [hep-ph]
18. M. Jarvinen, JHEP **1507**, 033 (2015). https://doi.org/10.1007/JHEP07(2015)033, arXiv:1501.07272 [hep-ph]

Chapter 8
Improved Holographic QCD with Magnetic Field

Electromagnetic properties of quarks and gluons are rich and complex, constituting an active branch of high energy research today. In this section we study the influence of external magnetic fields both on the ground and the plasma states from the holographic point of view. For reviews focusing more directly on the electromagnetic properties of QCD see e.g. [1, 2] and the references therein.

The problems we study here are relevant for experiment. In particular, intense magnetic fields are realized in off-central heavy-ion experiments, see Fig. 1.2. Such magnetic fields are comparable to the QCD scale Λ_{QCD} hence expected to influence the charge transport in the plasma substantially [3–10], see Fig. 8.1. Their rapid decay in time, together with expansion of the plasma in the presence of an external magnetic field, cause charge transport in the fluid due to ordinary electrodynamic phenomena such as the Faraday's law and the Lorentz force [10, 11]. RHIC isobar experiment [12] (see also [13]), whose data will become public in near future, particularly focuses on the effects of magnetic fields on transport in the QGP. Moreover, future-planned heavy-ion colliders FAIR and NICA, as well as the recently upgraded ALICE detector at CERN are expected provide new data on the electromagnetic properties of QCD. Finally, neutron stars and magnetars provide natural laboratories for studying cold, high density and magnetized quark-gluon matter.

Charge transport due to more subtle, quantum mechanical effects, such as the chiral magnetic effect [3, 5, 6, 14] and chiral magnetic wave [15] are in the focus of active research, on which, as we will see below, the holographic correspondence offers an alternative new angle. Finally, electromagnetic properties of strong interactions might also be crucial to the physics of the early universe [16].

Below we first focus on the dependence of the ground state, i.e. the quark condensate on the magnetic field. We will consider the phenomena of (inverse) magnetic catalysis and how to analyze these phenomena using holography. We then explore how magnetic fields affect the phase diagram of the quark-gluon matter. Finally, we compute the shear viscosity in holographic QCD in the presence of magnetic fields.

© The Author(s), under exclusive license to Springer Nature Switzerland AG 2021
U. Gürsoy, *Holography and Magnetically Induced Phenomena in QCD*,
SpringerBriefs in Physics,
https://doi.org/10.1007/978-3-030-79599-3_8

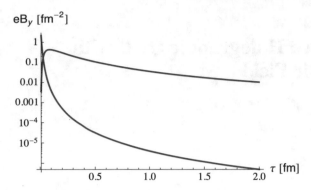

Fig. 8.1 Magnetic field at the origin in the center of mass frame as a function of proper time resulting from spectator and participant ions in an off-central heavy ion collisions at LHC with impact parameter $b = 7$ fm. Blue (red) curve is for electric conductivity $\sigma = 0$ ($\sigma = 0.023\ fm^{-1}$) respectively. Figure is from [10] and reprinted with permission from APS publishers

8.1   Background with Finite Magnetic Field and Temperature

For simplicity, we consider a constant magnetic field in one direction which we take x_3. Before we write down the holographic description of electromagnetic properties, we need to discuss following two issues. First, magnetic field couples directly only to quarks in the QGP. Hence, for the same reasons as explained in the previous chapter, its effects would be negligible in the large N QCD limit unless we also take $N_f \to \infty$ keeping the ratio fixed as in (7.1). Second, a general open problem of gauge-gravity duality is the lack of a clear description of U(1) gauge symmetry of the boundary field theory in the bulk. Instead, assuming that electromagnetic field is sourced externally, one replaces the desired U(1) electromagnetic symmetry by a global U(1) symmetry in the holographic description, whose description in the bulk becomes clear: it corresponds to a U(1) gauge field in the bulk.[1] For our discussion here treating U(1) as a global symmetry will be adequate as we are interested in its effects on the quark-gluon system as an external, non-dynamical field.

Accordingly, we will couple the magnetic field to the flavor sector of the holographic action as follows

$$S_f = -x\, M^3 N_c^2 \int d^5x\, V_f(\lambda, \tau) \sqrt{-\det\left(g_{\mu\nu} + w(\lambda)\, F_{\mu\nu}^V + \kappa(\lambda)\, \partial_\mu \tau\, \partial_\nu \tau\right)}. \quad (8.1)$$

---

[1] See [17, 18] for two proposals to resolve this open problem. The former proposes to describe the U(1) electromagnetic symmetry in terms of the center of the SU(N) gauge symmetry on the D-branes in the large N limit and identifies the dual bulk field as the Kalb–Ramond two-form of string theory. The latter arrives at a similar conclusion—that the bulk gauge field should be a two-form tensor from a different point of view based on holographic description of boundary one-form symmetries.

The potentials w, κ and V_f are given in Appendix A.2. We take the $U(1)_V$ bulk gauge field as

$$A_\mu^V = \left(A_t(r), -\frac{x_2 B}{2}, \frac{x_1 B}{2}, 0, 0 \right). \tag{8.2}$$

Temporal component corresponds to the quark chemical potential μ as in (7.9). When we discuss the numerical results below, we first will set $\mu = 0$ focusing on the effects of B only, then turn it on and explore the effects of both. The choice of spatial components in (8.2) corresponds to a constant magnetic field in the x_3 direction on the boundary at $r = 0$. As a result, the $SO(3)$ rotational symmetry of the plasma is broken down to a $O(2)$ subgroup around x_3 direction. Accordingly, we should modify the ansatz for the black-brane metric as

$$ds^2 = e^{2A(r)} \left(-f(r)dt^2 + dx_1^2 + dx_2^2 + e^{2W(r)}dx_3^2 + f(r)^{-1}dr^2 \right), \tag{8.3}$$
$$\phi = \Phi(r), \qquad \tau = \tau(r)$$

where we implemented breaking of rotational symmetry by the metric function $W(r)$. As before, there is a horizon at $r = r_h$ where f vanishes and one must require the same boundary asymptotics at $r \to 0$ as in the previous sections. In particular the new function $W \to 0$ as $r \to 0$. Finally, when comparing $B \neq 0$ and $B = 0$, the corresponding solutions should have the same value of the integration constants T, μ and Λ_{QCD}.

The ansatz (8.2) solves bulk Maxwell's equations automatically. One is left with the coupled non-linear system of Einstein's equations for the functions A, f and W, the dilaton equation of motion for ϕ and the tachyon equation of motion for τ. We present the equations of motion in Appendix A.3. Solving this system is not a straightforward exercise, yet manageable by a numerical code [19–21]. We will present the results in the next section.

Let us discuss the parameters of the model. The full system consists of second order equations for A, f, W, τ and $A_t(r)$ and a first order constraint equation[2] for Φ. This means 11 integration constants to fix. Inspecting these equations at the horizon fixes three of these constants: We require the gauge field A_t vanish at the horizon relating q to μ. In addition, combinations of (A.15), (A.16) and the constraint equation (A.19) on the horizon determine $W'(r_h)$ and $A'(r_h)$ in terms of other metric functions. Finally, the near horizon limit of the tachyon equation (A.21) fixes τ' in terms of $\tau(r_h)$, $\Phi(r_h)$ and $A(r_h)$. Therefore, we are left with 7 integration constants. Now we require that the background functions acquire fixed values on the boundary at the UV cut-off r_c near $r = 0$. One has to require that $f(r_c) = 1$ and $W(r_c) = 0$. These reduce the number of integration constants to 5.

These 5 integration constants correspond to physical parameters of the system: $A(r_c)$ determines the volume of the system. Since we deal with an extensive system this volume term will be a trivial factor that multiplies all thermodynamic quantities and drops out of dimensionless quantities. $A_t(u_c)$ corresponds to the chemical poten-

---

[2] The second order dilaton EoM can be derived from this constraint equation.

tial μ. The integration constant associated with Φ corresponds to Λ_{QCD} as before. The integration constant associated with f (that can be thought of as $f'(u_h)$) corresponds to T and finally the remaining integration constant associated to τ corresponds to the mass of the fermions. In addition, there is B which is constant throughout the solution. Finally, there are two more parameters in the action which we will vary below: the flavor to color ratio x and a constant c which parametrizes the function $w(\Phi)$, see Appendix A.2.[3] All the other functions in the flavor and glue actions are fixed as explained above. In this section we set the baryon chemical potential as well as the quark mass to zero: $\mu = m_q = 0$. This leaves us with the following free parameters: T, B, x and the parameter c.

A generic consequence of a magnetic field in the holographic background is presence of an (approximate) $AdS_3 \times R^2$ region in the deep interior of the geometry. This is expected by symmetry as the magnetic field becomes dominant in the IR, breaking AdS_5 UV asymptotics. Boosts along the direction of B remains intact. Roughly speaking, this direction combines with time and the holographic direction to form an AdS_3 and the two directions perpendicular to B form an R^2. This was first observed in [22] in a simple model of magnetic holography where the magnetic field was introduced via 5D Maxwell field. This trend has been continued in [23] which added a scalar field to this set-up, breaking the dual conformal symmetry, realizing a situation closer to real QGP. The same symmetry breaking pattern $AdS_5 \rightarrow AdS_3 \times R^2$ also arise in the more complicated case of the Veneziano limit considered in this review, not in the deep interior but in an intermediate region [24].

8.2   Quark-Antiquark Potential in a Magnetized State

Before discussing the effects of B on the phase diagram we consider its influence on another important observable, the quark-antiquark potential. Lattice results are available for this observable in the presence of a magnetic field [25]. Below we present the result from holography, obtained in [24]. Comparison to lattice then determines the preferred value for c, which turns out to be between $c = 0.25$ and 0.4 with a slightly better fit for 0.25.[4]

Even though the presence of a magnetic field induces an intermediate $AdS_3 \times \mathbb{R}^2$ region [24], the effect of B on the quark-antiquark potential is qualitatively the same as before. In particular B does not destroy linear confinement. The reason is that, in the V-QCD setup B enters through the flavor sector, in contrast to Einstein–Maxwell models [22, 23]. Since, in our model, the flavor sector decouples in the deep IR, the effect of B on the geometry is negligible, and far IR geometry is the same as when $B = 0$. As a consequence of this, linear confinement is not lost in the presence of B.

---

[3] As clear from dependence of w on c in Appendix A.2, this parameter controls dependence of the flavor sector on the 't Hooft coupling λ.

[4] Determination of this quantity more precisely requires a full-fledged global analysis of all observables affected by B.

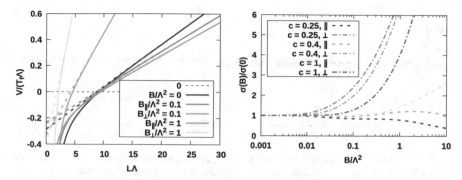

Fig. 8.2 Quark-antiquark potentials and string tensions for $a = 0$. The plots present two cases, corresponding to the $q\bar{q}$ pair separated along the direction of the magnetic field ($B_\parallel$) or one of the directions orthogonal to it ($B_\perp$), all in units of $\Lambda \sim 1$ GeV. The potentials themselves (left panel) are shown together with the asymptotic behavior of the potentials for large separation (shown dashed). The string tensions (right panel) represent the slope of the potentials in the confining regime. For these plots we varied the parameter c introduced in (A.13), which controls the region of intermediate energy scales. The $c = 0.25$ case displays both IMC and agrees qualitatively with the results of [25], in the range of magnetic fields considered there $eB \sim 0 - 1.2$ GeV. Figure from [24]

The computation follows the same steps presented in Sect. 5.3, see [24] for details. In Fig. 8.2 (left panel) we have plotted the quark-antiquark potential for two cases, labeled as $B_\parallel$ and $B_\perp$. The first one corresponds to separation of the quark-antiquark pair in the same direction as the magnetic field, i.e., V_3, while the second one corresponds to separation orthogonal to the magnetic field, i.e., $V_{1,2}$, both for $c = 0.25$. It is interesting to compare the string tensions (slopes) of this confining regime with lattice QCD calculations [25], at least qualitatively. In Fig. 8.2 (right panel) we plot such tensions (normalized by the $B = 0$ value) for different choices of c, observing that the $c = 0.25$ case agrees better with [25] in the sense the latter also finds monotonically increasing/decreasing functions of B for the perpendicular/parallel cases, respectively for $eB \sim 0 - 1.2$ GeV.

8.3 Phase Diagram of ihQCD with Magnetic Field

The phase diagram following from this holographic model has been studied in [21]. Coupling of B to the background is controlled by the ratio of flavors to glue x that sits in front of the flavor action (8.1) and the function w that enters in this action. The result is qualitatively similar to the one at finite T and μ, see Fig. 7.1. Generically, there are three phases: confined—chiral symmetry broken, deconfined—chiral symmetry broken and deconfined—chiral symmetry restored. The first two are separated by a first order deconfinement transition line that we denote by $T_d(B)$. The last two are separated by a second order transition line that we denote by $T_\chi(B)$. We plot these functions in Fig. 8.3 for $x = 1$ and various choices of c.

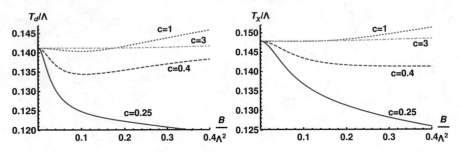

Fig. 8.3 Deconfinement (left) and the chiral symmetry restoration (right) temperatures as a function of B for different choices of the parameter c which parametrizes response of background to magnetic field in ihQCD. Figures reproduced from [21]

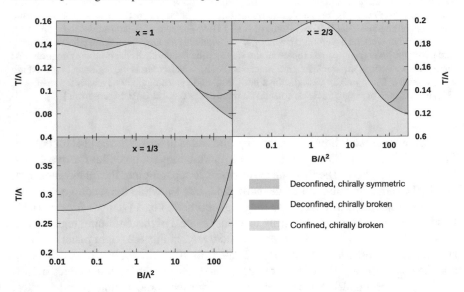

Fig. 8.4 Phase diagram of ihQCD in the presence of an external magnetic field for various values of the ratio $x = N_f/N_c$. Plots adapted from [21]

It turns out that the smaller values $c = 0.25 - 0.4$ gives better qualitative agreement with recent lattice QCD results in [26–29]. In particular the phenomenon of "inverse magnetic catalysis" observed on the lattice is also present in the holographic model for this choice. We will assume either of $c = 0.4$ or $c = 0.25$ below. We observe that, for small c, both T_d and T_χ has a non-trivial profile, in particular they exhibit a minimum in B. The decrease at small B is associated with the phenomenon of inverse magnetic catalysis which we discuss in the next Sect. 8.4.

One also finds that the phase diagram undergoes non-trivial changes when the flavor-to-color ratio x is varied. In Fig. 8.4 we show the diagram for the choice for the various values of x. We observe that the deconfined/chirally broken phase is more pronounced for larger values of x. In fact, this phase disappears altogether for $x \ll 1$

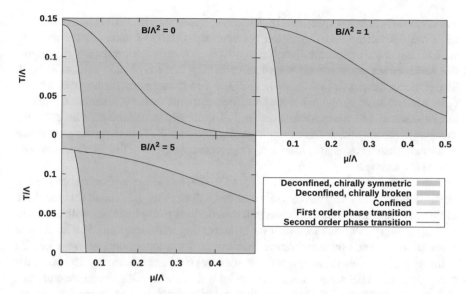

Fig. 8.5 Phase diagram of the ihQCD theory in the Veneziano limit with flavor-to-color ratio $x = 1$ and with finite quark chemical potential and for various magnetic fields. Figure from [30]

as shown in [21]. It exists for $x \gtrsim 1/3$. Finally for even larger values $x \approx 1$ the same phase also appears also for small values of B, in agreement with the $\mu \to 0$ edge of the phase diagram in Fig. 7.1.

We show the phase diagram on the $T - \mu$ plane for the various choices of the magnetic field in Fig. 8.5, including vanishing B (same as Fig. 7.1) on the top left corner. We observe that the chirally broken plasma phase (shown by blue in the figure) is manifest at finite μ and B. This is one of the generic predictions of the model. This phase also exists for sufficiently large magnetic fields where the condensate is strengthened rather than weakened by magnetic fields, a characteristic of magnetic catalysis. This finding is consistent with the perturbative QCD result valid for $eB \gg \Lambda^2_{QCD}$ [1]. Finally, we observe that the first order deconfinement transition merges with the chiral symmetry restoration transition at finite μ and large B. Close inspection of the merger unveils a bifurcation of the first and the second order transition boundaries, and that the solid line ends on a separate critical point [30].

8.4 Inverse Magnetic Catalysis

It is known from studies based on perturbative QCD and effective field theory that the quark condensate is strengthened in the presence of a magnetic field; a phenomenon called "magnetic catalysis" [31–33]. One can qualitatively understand the reason behind this phenomenon as follows. Turning on a magnetic field results in Landau quantization in fermion spectrum. In particular, momenta in the directions transverse to B are discretized and separation between the discrete Landau states increase with B. Landau quantization therefore restricts motion in the transverse directions. For large values of B ground state has vanishing transverse momentum. This, in turn projects the physics of flavor in QCD to 1+1 dimensions for large B. On the other hand, it is well-known that the IR effects responsible for condensate formation are stronger in 1+1 dimensions than in 3+1 dimensions, resulting in an effective increase in the magnitude of the quark condensate with B. This suggestive argument is substantiated by explicit calculations in perturbative QCD in the references above, see [1] for a review.

Lattice studies of QCD with 2+1 flavors [26–29] reveal a complicated behavior at strong coupling for energy scales relevant to QGP physics. It is found that for finite but small temperatures, the condensate in the confined phase increases with B up to a certain turning point, above which it starts decreasing with increasing B. This critical value of B depends on temperature. Moreover, for temperatures above a certain value, slightly below the deconfinement crossover temperature, around 150 MeV, the condensate starts decreasing even for smaller B down to $B = 0$. Therefore one finds that strong coupling in QCD triggers the opposite effect, coined "inverse magnetic catalysis" (IMC).

The precise physical mechanism for this behavior has not yet been clarified. There are indications from further lattice studies [34, 35] that the presence of a turning point in the condensate as a function of B results from a competition between two separate sources. Considering the path integral $\langle \bar{q}q \rangle$ one can identify these two sources as follows. First, there is a direct coupling to B of the fermion propagators inside the operator $\bar{q}q$ in the path integral. This source is called the "valence quarks" in [34] and it always tends to strengthen the condensate, essentially for the same reason explained above that leads to magnetic catalysis. There is a second source of coupling to B however, that comes from the quark determinant arising from the gluon path integral. This second source, called "sea quarks" is weak at weak coupling compared to former, hence it can be neglected, and one finds magnetic catalysis. However, it becomes stronger at intermediate or large values of the coupling constant, and it was argued in [34, 35] that it dominates over the first source for relatively large values of B and T, leading to the inverse effect. See [36] for a similar suggestion where the authors propose that IMC results from a combined effect of gluon screening and the weakening of gauge coupling at high energies. These are mostly suggestive arguments and it is desirable to investigate the question using an alternative approach such as holographic QCD.

The question has been addressed in holography in various works [37–45] for toy systems involving adjoint flavors or small number of fundamental quarks for which the fermion contribution to the background is suppressed at large N. Recently, the question has been addressed in [21] for ihQCD in the Veneziano limit (7.1), where it was found that holography supports the valence versus sea quark suggestion of [34, 35].

In Fig. 8.6 we show the phase transition boundaries for the deconfinement and chiral symmetry restoration transitions for $x = 1$ and $c = 0.4$. We observe that both of

Fig. 8.6 Phase diagram and curves of constant $\langle \bar{q}q \rangle$ in ihQCD for a choice of $x = 1$ and $c = 0.4$. Dimensionful quantities are normalized with the integration constant Λ of ihQCD that is proportional to the in intrinsic energy scale of QCD Λ_{QCD}. Plot reproduced from paper [21]

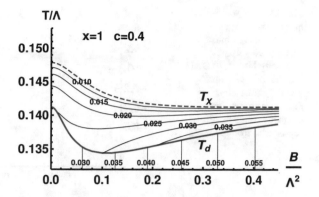

these transition temperatures initially decrease with increasing B. In the deconfined—chiral symmetry broken phase, $T_d < T < T_\chi$, this means that it becomes easier to melt the condensate with larger B. We also show contours of constant condensate in the same plot. This provides a direct confirmation that the condensate decreases with B at least for small enough B. One observes that the curves of constant condensate extend between the curves $T_d(B)$ and $T_\chi(B)$ continuously decreasing with increasing T and finally vanishing at T_χ leading to the second order chiral symmetry restoration transition discussed in the previous section.[5]

Finally in Fig. 8.7 we plot the renormalization invariant and dimensionless combination $\Delta\Sigma(T, B) = \Sigma(T, B) - \Sigma(T, 0)$ where

$$\Sigma(T, B) = \frac{\langle \bar{q}q \rangle(T, B)}{\langle \bar{q}q \rangle(0, 0)} = \frac{1}{\langle \bar{q}q \rangle(0, 0)} \left(\langle \bar{q}q \rangle(T, B) - \langle \bar{q}q \rangle(0, 0) \right) + 1. \quad (8.4)$$

We observe that, in qualitative agreement with the lattice results described above, the condensate increases with B up to a certain value of the temperature around $T/\Lambda \approx 0.138$, and it starts decreasing for larger T up to the chiral symmetry restoration transition. For larger T the condensate drops to zero for large B, as demonstrated by the blue curve in Fig. 8.7, because the condensates vanishes above T_χ, cf. Fig. 8.6.

The suggestion of [34, 35] for the physical mechanism behind the inverse magnetic catalysis relating it to the "sea quarks" as described above can be tested in holography. The two sources of coupling of the condensate to B, the direct coupling called the valence quarks, and the indirect, glue induced coupling called the sea quarks can be identified in holography with two analogous sources as follows: The condensate is determined by solving the tachyon equation motion. This equation depends on B again in two separate ways. First, an explicit dependence, which we identify with the valence quarks; second an indirect dependence arising from dependence

---

[5] Vertical contours below T_d if an artefact of holographic QCD: temperature dependence in the confined phase, that corresponds to the thermal gas solution (6.3) is suppressed in the large-N limit. To capture this dependence one needs fluctuations of the background fields, leading to a $\mathcal{O}(1/N)$ corrections of the free energy.

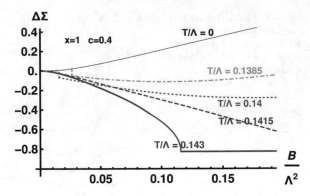

Fig. 8.7 The normalized quark condensate as a function of B in the deconfined—chiral symmetry broken phase in ihQCD for $x = 1$ and $c = 0.4$ shows clear demonstration of inverse magnetic catalysis. Plot reproduced from paper [21]

of background functions on B, which we can identify with the sea quarks. Various tests of this suggestion is made in [21] by isolating either of the two dependences by playing with the values of B and x and strong indications found supporting this suggestion.

What happens to the phenomenon of inverse magnetic catalysis at finite quark density? This question cannot be easily addressed on the lattice due to the infamous sign problem [46]. Turning on both a finite chemical potential and magnetic field in the holographic setting, the question was investigated in [30].

Figure 8.8 (left) shows that the chiral restoration temperature decreases (increases) for small (large) quark chemical potential μ signaling IMC for small potential. The regime IMC is observed in the temperature/chemical potential plane is shown by red in Fig. 8.8 (right).

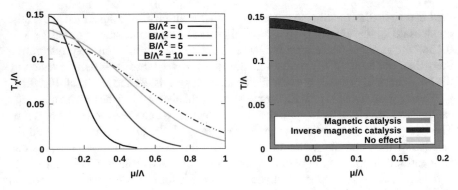

Fig. 8.8 (Left) Chiral symmetry restoration temperature as a function of quark chemical potential for various values of B. (Right) Presence of inverse magnetic catalysis in improved holographic QCD at finite quark chemical potential

8.5 Speed of Sound

Sound waves in the plasma are pressure waves of alternating high and low pressure regions along the wave motion. In the absence of magnetic field and chemical potential, the typical speed of sound waves are determined by how pressure density varies with the energy density,

$$c_s^2 = \frac{dp}{d\epsilon} = \frac{sdT}{Tds},$$ (8.5)

where we used the first law of thermodynamics in the second equation. At finite μ and B, this is generalized to

$$c_s^2 = \left.\frac{sdT + nd\mu}{Tds + \mu dn + BdM}\right|_{n/s,B},$$ (8.6)

where n is, quark density and M is magnetization.

Sound speed is an essential quantity which characterizes the equation of state of the plasma. Hence, constraints on its value arising from microscopic physics would be extremely interesting. Such constraints are implied by holography at strong coupling. Indeed, [47, 48] showed that the speed of sound approaches to its conformal value $c_s^2 = 1/3$ *from below* at high temperatures, when conformality is broken by a single relevant scalar operator. The latter is also the approximation that we employ

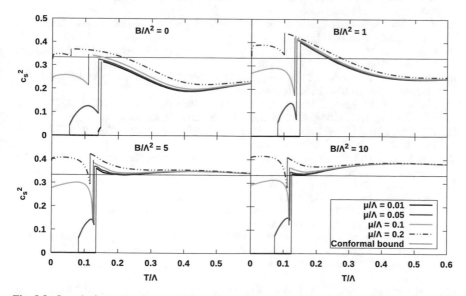

Fig. 8.9 Speed of sound in holographic QCD as a function of temperature for different values of chemical potential and magnetic field. The dashed horizontal line corresponds to the conformal value $c_s^2 = 1/3$

in improved holographic QCD. We are, then, curious to know whether similar holographic bounds can be placed at finite quark density or magnetic field. The answer is unknown. However, one can show, by direct computation [30], that the proposed conformal bound is clearly violated in these more complicated cases as we show in Fig. 8.9. Whether, there is another universal upper bound (below the speed of light) at strong coupling, also at finite μ and B remains to be seen.

References

1. V.A. Miransky, I.A. Shovkovy, Phys. Rep. **576**, 1 (2015). https://doi.org/10.1016/j.physrep. 2015.02.003, arXiv:1503.00732 [hep-ph]
2. D.E. Kharzeev, K. Landsteiner, A. Schmitt, H.U. Yee, Lect. Notes Phys. **871**, 1 (2013). (arXiv:1211.6245 [hep-ph])
3. D.E. Kharzeev, L.D. McLerran, H.J. Warringa, Nucl. Phys. A **803**, 227 (2008). https://doi.org/ 10.1016/j.nuclphysa.2008.02.298 (arXiv:0711.0950 [hep-ph])
4. V. Skokov, A.Y. Illarionov, V. Toneev, Int. J. Mod. Phys. A **24**, 5925 (2009). https://doi.org/ 10.1142/S0217751X09047570, arXiv:0907.1396 [nucl-th]
5. K. Tuchin, Phys. Rev. C **82**, 034904 (2010). Erratum: [Phys. Rev. C **83** (2011) 039903]. https://doi.org/10.1103/PhysRevC.83.039903, https://doi.org/10.1103/PhysRevC.82.034904, arXiv:1006.3051 [nucl-th]
6. V. Voronyuk, V.D. Toneev, W. Cassing, E.L. Bratkovskaya, V.P. Konchakovski, S.A. Voloshin, Phys. Rev. C **83**, 054911 (2011). https://doi.org/10.1103/PhysRevC.83.054911, arXiv:1103.4239 [nucl-th]
7. W.T. Deng, X.G. Huang, Phys. Rev. C **85**, 044907 (2012). https://doi.org/10.1103/PhysRevC. 85.044907, arXiv:1201.5108 [nucl-th]
8. K. Tuchin, Adv. High Energy Phys. **2013**, 490495 (2013). https://doi.org/10.1155/2013/ 490495, arXiv:1301.0099 [hep-ph]
9. L. McLerran, V. Skokov, Nucl. Phys. A **929**, 184 (2014). https://doi.org/10.1016/j.nuclphysa. 2014.05.008, arXiv:1305.0774 [hep-ph]
10. U. Gursoy, D. Kharzeev, K. Rajagopal, Phys. Rev. C **89**(5), 054905 (2014). https://doi.org/10. 1103/PhysRevC.89.054905, arXiv:1401.3805 [hep-ph]
11. U. Gürsoy, D. Kharzeev, E. Marcus, K. Rajagopal, C. Shen, Phys. Rev. C **98**(5), 055201 (2018). https://doi.org/10.1103/PhysRevC.98.055201, arXiv:1806.05288 [hep-ph]
12. P. Tribedy [STAR], J. Phys. Conf. Ser. **1602**(1), 1 (2020). https://doi.org/10.1088/1742-6596/ 1602/1/012002, arXiv:2009.01230 [nucl-ex]
13. D.E. Kharzeev, J. Liao, Nature Rev. Phys. **3**(1), 55–63 (2021). https://doi.org/10.1038/s42254-020-00254-6, arXiv:2102.06623 [hep-ph]
14. K. Fukushima, D.E. Kharzeev, H.J. Warringa, Phys. Rev. D **78**, 074033 (2008). https://doi.org/ 10.1103/PhysRevD.78.074033, arXiv:0808.3382 [hep-ph]
15. D.E. Kharzeev, H.U. Yee, "Chiral Magnetic Wave," Phys. Rev. D **83**, 085007 (2011). https:// doi.org/10.1103/PhysRevD.83.085007
16. K. Subramanian, Rep. Prog. Phys. **79**(7), 076901 (2016). https://doi.org/10.1088/0034-4885/ 79/7/076901, arXiv:1504.02311 [astro-ph.CO]
17. U. Gursoy, JHEP **12**, 062 (2010). https://doi.org/10.1007/JHEP12(2010)062, arXiv:1007.4854 [hep-th]
18. D.M. Hofman, N. Iqbal, SciPost Phys. **4**(1), 005 (2018). https://doi.org/10.21468/SciPostPhys. 4.1.005, arXiv:1707.08577 [hep-th]
19. T. Drwenski, U. Gursoy, I. Iatrakis, JHEP **12**, 049 (2016). https://doi.org/10.1007/ JHEP12(2016)049, arXiv:1506.01350 [hep-th]
20. T. Demircik, U. Gursoy, arXiv:1605.08118 [hep-th]

21. U. Gürsoy, I. Iatrakis, M. Järvinen, G. Nijs, arXiv:1611.06339 [hep-th]
22. E. D'Hoker, P. Kraus, Lect. Notes Phys. **871**, 469–502 (2013). https://doi.org/10.1007/978-3-642-37305-3_18, arXiv:1208.1925 [hep-th]
23. R. Rougemont, R. Critelli, J. Noronha, Phys. Rev. D **91**(6), 066001 (2015). https://doi.org/10.1103/PhysRevD.91.066001, arXiv:1409.0556 [hep-th]
24. U. Gürsoy, M. Järvinen, G. Nijs, J.F. Pedraza, JHEP **03**, 180 (2021). https://doi.org/10.1007/JHEP03(2021)180, arXiv:2011.09474 [hep-th]
25. C. Bonati, M. D'Elia, M. Mariti, M. Mesiti, F. Negro, F. Sanfilippo, Phys. Rev. D **89**(11), 114502 (2014). https://doi.org/10.1103/PhysRevD.89.114502, arXiv:1403.6094 [hep-lat]
26. G.S. Bali, F. Bruckmann, G. Endrodi, Z. Fodor, S.D. Katz, S. Krieg, A. Schafer, K.K. Szabo, JHEP **1202**, 044 (2012). https://doi.org/10.1007/JHEP02(2012)044 (arXiv:1111.4956 [hep-lat])
27. G.S. Bali, F. Bruckmann, G. Endrodi, Z. Fodor, S.D. Katz, S. Krieg, A. Schafer, K.K. Szabo, PoS LATTICE **2011**, 192 (2011). arXiv:1111.5155 [hep-lat]
28. G.S. Bali, F. Bruckmann, G. Endrodi, Z. Fodor, S.D. Katz, A. Schafer, Phys. Rev. D **86**, 071502 (2012). https://doi.org/10.1103/PhysRevD.86.071502, arXiv:1206.4205 [hep-lat]
29. M. D'Elia, Lect. Notes Phys. **871**, 181 (2013). (arXiv:1209.0374 [hep-lat])
30. U. Gursoy, M. Jarvinen, G. Nijs, Phys. Rev. Lett. **120**(24), 242002 (2018). https://doi.org/10.1103/PhysRevLett.120.242002, arXiv:1707.00872 [hep-th]
31. V.P. Gusynin, V.A. Miransky, I.A. Shovkovy, Phys. Rev. Lett. **73**, 3499 (1994). Erratum: [Phys. Rev. Lett. **76**, 1005 (1996)]. https://doi.org/10.1103/PhysRevLett.73.3499 [hep-ph/9405262]
32. V.P. Gusynin, V.A. Miransky, I.A. Shovkovy, Phys. Lett. B **349**, 477 (1995). https://doi.org/10.1016/0370-2693(95)00232-A, arXiv:hep-ph/9412257
33. V.P. Gusynin, V.A. Miransky, I.A. Shovkovy, Phys. Rev. D **52**, 4718 (1995). https://doi.org/10.1103/PhysRevD.52.4718 (arXiv:hep-th/9407168)
34. F. Bruckmann, G. Endrodi, T.G. Kovacs, JHEP **1304**, 112 (2013). https://doi.org/10.1007/JHEP04(2013)112, arXiv:1303.3972 [hep-lat]
35. F. Bruckmann, G. Endrodi, T.G. Kovacs, arXiv:1311.3178 [hep-lat]
36. N. Mueller, J.M. Pawlowski, Phys. Rev. D **91**(11), 116010 (2015). https://doi.org/10.1103/PhysRevD.91.116010, arXiv:1502.08011 [hep-ph]
37. A.V. Zayakin, JHEP **0807**, 116 (2008). https://doi.org/10.1088/1126-6708/2008/07/116 (arXiv:0807.2917 [hep-th])
38. V.G. Filev, D. Zoakos, JHEP **1108**, 022 (2011). https://doi.org/10.1007/JHEP08(2011)022, arXiv:1106.1330 [hep-th]
39. J. Erdmenger, V.G. Filev, D. Zoakos, JHEP **1208**, 004 (2012). https://doi.org/10.1007/JHEP08(2012)004 (arXiv:1112.4807 [hep-th])
40. F. Preis, A. Rebhan, A. Schmitt, Lect. Notes Phys. **871**, 51 (2013). (arXiv:1208.0536 [hep-ph])
41. K.A. Mamo, JHEP **1505**, 121 (2015). https://doi.org/10.1007/JHEP05(2015)121 (arXiv:1501.03262 [hep-th])
42. D. Dudal, D.R. Granado, T.G. Mertens, Phys. Rev. D **93**(12), 125004 (2016). https://doi.org/10.1103/PhysRevD.93.125004, arXiv:1511.04042 [hep-th]
43. R. Rougemont, R. Critelli, J. Noronha, Phys. Rev. D **93**(4), 045013 (2016). https://doi.org/10.1103/PhysRevD.93.045013, arXiv:1505.07894 [hep-th]
44. N. Evans, C. Miller, M. Scott, Phys. Rev. D **94**(7), 074034 (2016). https://doi.org/10.1103/PhysRevD.94.074034, arXiv:1604.06307 [hep-ph]
45. N. Jokela, A.V. Ramallo, D. Zoakos, JHEP **1402**, 021 (2014). https://doi.org/10.1007/JHEP02(2014)021, arXiv:1311.6265 [hep-th]
46. P. de Forcrand, PoS **LAT2009**, 010 (2009). https://doi.org/10.22323/1.091.0010, arXiv:1005.0539 [hep-lat]
47. P.M. Hohler, M.A. Stephanov, Phys. Rev. D **80**, 066002 (2009). https://doi.org/10.1103/PhysRevD.80.066002, arXiv:0905.0900 [hep-th]
48. A. Cherman, T.D. Cohen, A. Nellore, Phys. Rev. D **80**, 066003 (2009). https://doi.org/10.1103/PhysRevD.80.066003, arXiv:0905.0903 [hep-th]

Chapter 9
Hydrodynamics and Transport Coefficients

One of the major applications of the holographic correspondence in strongly inter-acting plasmas is the characterization of transport of conserved quantities, such as energy-momentum, charge and angular momentum. Magnitude of these currents are determined by the associated transport coefficients that quantify dissipation in the system, such as shear and bulk viscosities, and conductivity. In QFT, these quantities are obtained from retarded Green's functions of operator that characterize transport, e.g. the energy-momentum tensor and electric current, using Kubo's linear response theory [1]. At strong coupling, holography becomes a major tool to compute these transport coefficients, when it is applicable. Below, we first focus on computation of viscosities in the quark-gluon plasma using the ihQCD theory and then discuss a relatively less understood type of transport, anomalous transport, that is particularly important in the presence of intense magnetic fields, see Chap. 1.

9.1 Shear and Bulk Viscosity

Hydrodynamic gradient expansion was introduced and two of the first order transport coefficients, shear and bulk viscosity were discussed in Chap. 3. Furthermore, in Sect. 4.2 we showed that, in the absence of string corrections to Einstein's gravity, holography yields a universal value for the ratio of shear viscosity to entropy, that is $\eta/s = 1/4\pi$. Thus, we do not need to compute it again in ihQCD model as its value is guaranteed to remain unchanged.[1]

---

[1] This changes in an anisotropic state, e.g. in the presence of an external magnetic field as we discuss below.

© The Author(s), under exclusive license to Springer Nature Switzerland AG 2021
U. Gürsoy, *Holography and Magnetically Induced Phenomena in QCD*,
SpringerBriefs in Physics,
https://doi.org/10.1007/978-3-030-79599-3_9

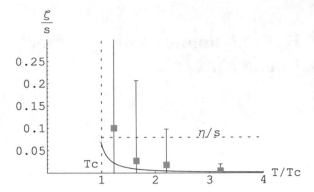

Fig. 9.1 Result of the ihQCD calculation for the bulk viscosity compared with the lattice data of [3]

Next, we turn to bulk viscosity. This is the coefficient that characterizes response of the system to a volume deformation. From Eq. (3.8), we learn that it is determined from the computation of the ii component of the stress tensor two-point function. The corresponding fluctuation equation in holography is given by

$$\ddot{\delta g}_{ii} + \dot{\delta g}_{ii}\left(3\dot{A} + \frac{\dot{f}}{f} + 2\frac{\dot{X}}{X}\right) + \left(\frac{\omega^2}{f} - \frac{\dot{f}}{f}\frac{\dot{X}}{X}\right)\delta g_{ii} = 0. \qquad (9.1)$$

This equation does not exhibit any universality at the horizon, because of the presence non-vanishing mass term in the limit $\omega \to 0$ and the result, that is a non-trivial function of T, indeed depends on the choice of the potential in (5.1). For the choice (5.43) ihQCD theory gives [2] the plot given in Fig. 9.1. We observe two features in Fig. 9.1. First, the bulk viscosity increases towards the deconfinement transition at $T = T_c$. Second, the ratio ζ/s vanishes at very large temperatures, a result qualitatively consistent with perturbative QCD. In this plot we compare our result with the lattice QCD calculation of [3]. The latter calculation involves large systematic and statistical errors. These errors are due to the fact that, to obtain a real-time correlation function such as (3.7) from the lattice, one needs to analytically continue the Euclidean correlators, that necessitate the knowledge of the entire spectral density of QCD associated with the energy-momentum tensor [3], an information that we do not have. The ihQCD result quantitatively agrees with another holographic model for QCD [4]. It should also be compared with the result of the Bayesian analyses we presented in Fig. 3.2. We see that holography is qualitatively and, to a large extent quantitatively, consistent with other approaches.

9.2 Shear Viscosity in the Presence of Magnetic Field

We have seen in Sect. 4.2 that shear viscosity to entropy density takes the universal value of $\eta/s = 1/4\pi$ [5, 6] in to-derivative holography and is in remarkable agreement with experimental data [7]. In fact, this is only true in an isotropic state. In an

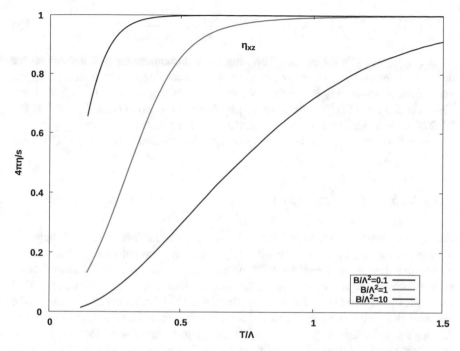

Fig. 9.2 Shear viscosity to entropy ratio of the longitudinal component $\eta_{xz} = \eta_{yz}$ as a function of temperature (in units of $\Lambda \sim 1$ GeV) for $c = 0.25$ in holographic QCD with a magnetic field along z. The curves are cut off at the chiral transition temperature below which there is a non-trivial chiral condensate

anisotropic situation, caused, for example, by presence of an external magnetic field or different pressure gradients in different directions—a situation generic to heavy ion collisions—the shear viscosity on the (xy), (xz) and (yz) planes could be all different [8–20]. In case of partial breaking of isotropy, $SO(3) \to SO(2)$, the shear viscosity/entropy ratio on the plane perpendicular to the anisotropy vector continues to assume its universal value $1/4\pi$ as the proof of universality presented in Sect. 4.2 holds. This is gives for example $\eta_{xy}/s = 1/4\pi$ in the presence B in the z-direction.

In general the shear viscosity tensor can be computed via the standard Kubo formula,

$$\eta_{ij} = -\frac{1}{\omega} \text{Im} \langle T_{ij}(\omega, \mathbf{k}_1) T_{ij}(\omega, \mathbf{k}_2) \rangle \big|_{\omega \to 0, \, \mathbf{k}_{1,2} \to 0}, \tag{9.2}$$

where the limit on the right is taken first. For theories that break rotations, there are various independent components of the shear viscosity tensor. Here, we are motivated by the influence of magnetic field in the QGP and we only this as the source of anisotropy.[2] In this case the metric is given by (8.3) and the resulting shear viscosities are given as (see [21] for a recent derivation)

---

[2] See [21] for a discussion of different sources of anisotropy in the holographic QCD.

$$\frac{\eta_{xy}}{s} = \frac{1}{4\pi}, \qquad \frac{\eta_{xz}}{s} = \frac{\eta_{yz}}{s} = \frac{e^{2W(r_h)}}{4\pi} \qquad (9.3)$$

In Fig. 9.2 we plot the results. The longitudinal components of the shear tensor decrease monotonically from the UV to the IR as also observed in other anisotropic backgrounds [13–16, 18, 20]. In the UV the universal value $\eta_{ij}/s = 1/4\pi$ is attained. This is because the 5D background, (8.3) is chosen to be asymptotically AdS. In the IR they attain a smaller, non-zero value. See [21] for more on interpretation of these results and implications for the quark-gluon plasma created in heavy ion collisions.

9.3 Anomalous Transport

Finally, we consider another type of phenomena, anomalous transport, that is believed to occur in the quark-gluon plasma produced in heavy ion collisions. This is dissipation free means of energy-momentum and charge transport in QCD and other systems where chiral fermions participate in transport such as the Dirac and Weyl semimetals,[3] induced by quantum anomalies of the axial current [23, 24] and in the presence of a parity even vector source such as the external magnetic field or vorticity. Among several comprehensive reviews of the subject, see [25–29] for an exposition from quantum field theory viewpoint, and [30, 31] for holographic description. In this section, we discuss a only a subset of anomalous transport phenomena that is relevant for the QGP physics: the *chiral magnetic effect* [25, 32, 33].

The axial current, J^5 in QCD is classically conserved in QCD with massless quarks. As we discussed in Chap. 2, this is violated at the quantum level due to quantum anomalies. The anomaly can be understood as arising from the so-called triangle Feynman diagrams with J^5 connecting one corner, two gauge fields connecting the other two corners and fermions running in the loop. This is presented in standard QFT books, see e.g. [34]. There are basically two contributions (in the absence of vortices): an electromagnetic anomaly determined by a $J_5 - A - A$ triangle diagram with A denoting electromagnetic gauge field, and a QCD anomaly determined by a $J_5 - G - G$ diagram with G denoting the gluon field [23, 24].:

$$J_5^{\mu} = \sum_{i=1}^{N_f} \bar{\psi}_i \gamma^{\mu} \gamma^5 \psi_i, \qquad \partial_{\mu} J_5^{\mu} = \epsilon_{\mu\nu\alpha\beta} \left(c_1 \, F_V^{\mu\nu} F_V^{\alpha\beta} + c_2 \, \text{tr} \left(F^{\mu\nu} F^{\alpha\beta} \right) \right). \quad (9.4)$$

Here J_5 is the axial current, and F_V and F denote the electromagnetic and gluon the field strengths respectively. c_1 and c_2 are the electromagnetic and QCD anomaly coefficients. The last term on the RHS of the second equation is a topologic invariant, non-vanishing in the presence of non-trivial gluon winding:

---

[3] See for example [22] for observation of anomalous transport in Dirac semimetals.

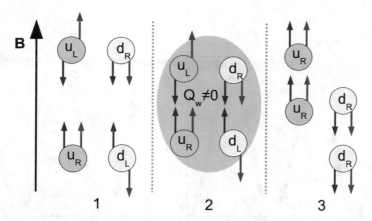

Fig. 9.3 Schematic description of the chiral magnetic effect. (Anti)-particle spins (denoted by blue arrows) are oriented (anti)-parallel to the magnetic field. In the chiral limit they move in the direction of their spin, or opposite to it, depending on whether they are left or right-handed. At time $t = 1$ both electric and chiral currents vanish. At $t = 2$ a gluon configuration with non-trivial winding is generated and this breaks conservation of chirality due to quantum anomaly. At $t = 3$ this non-trivial gluon configuration decays, producing an imbalance in chirality due to (9.4). This, eventually leads to an electric current. Plot is reprinted with permission from APS publishers [25]

$$Q_w = \frac{1}{24\pi^2} \epsilon_{\mu\nu\alpha\beta} \int d^4x \, \mathrm{tr} \left(F^{\mu\nu} F^{\alpha\beta} \right) \in \mathbb{Z}. \tag{9.5}$$

In QGP we are interested in temperatures much larger than physical quark masses $m_q/T \ll 1$, hence the axial current is indeed effectively conserved at the classical level. However, quantum anomalies (9.4) result in anomalous transport such as the chiral magnetic effect (CME). This is generation of electric current in the direction of external magnetic field.

The mechanism that leads to CME is schematically described in Fig. 9.3. Quark spins are aligned with B as a result of the Zeeman interaction. As quark masses can be neglected at high temperatures, in the absence of an external $\mathbf{E} \cdot \mathbf{B}$ term (first term in (9.4) and gluon fields with non-trivial topology (the second term in (9.4), axial charge is effectively conserved both at classical and quantum levels. This means helicity $h = \mathbf{S} \cdot \mathbf{p}/|\mathbf{p}|$, is approximately conserved and the quarks will move parallel or anti-parallel to B depending on their helicity eigenvalue. Because there are initially equal number of left and right handed particles then there is no net generation of electric current at time $t = 1$. Now suppose that a gluon configuration with a non-trivial topology is generated at $t = 2$ which decays at $t = 3$. This would then convert some of the left (right) movers into right (left) movers following the second term in the anomaly equation (9.4) resulting in chiral imbalance. Then, there is still effective conservation of the axial charge in phase $t = 3$ both at the classical and the quantum level since the non-trivial gluon configuration decayed, but now there is a net electric current in the direction of **B**. This current is given by

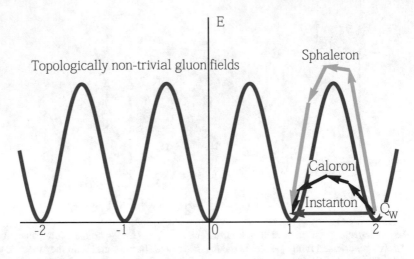

Fig. 9.4 Non-perturbative processes that change gluon winding number. Sphalerons are the unstable gluon field configurations sitting on top of the potential

$$\mathbf{J_V} = \sigma_B \mathbf{B} = c_1 \mu_5 \mathbf{B} \,, \qquad (9.6)$$

as shown both in field theory [33] and hydrodynamics [35]. Here an effective chemical potential μ_5 for the axial charge is introduced to implement its non-conservation in (9.4). For example this μ_5 will be non-zero if gluon configurations with non-trivial topology is generated at $t = 2$ of Fig. 9.3.

Among possible sources for generation of such non-trivial gluon configurations, the sphalerons [36], dominate at temperatures of order $T > \Lambda_{QCD}$ [37–39]. In Fig. 9.4 this mechanism is described schematically. In this figure we plot the vacuum energy of QCD as a function of the gluon-winding number Q_w (9.5). A sphaleron is an unstable field configuration, given by the maxima in this figure, produced at sufficiently high temperatures. They decay by thermal fluctuations and generate a net change in Q_w. In QCD probability of this process is quantified by a transport coefficient called the *sphaleron decay rate* (or Chern–Simons decay rate) [37–39]. Define the topological charge by,

$$q(x^\mu) \equiv \frac{1}{16\pi^2} \mathrm{tr}\,[F \wedge F] = \frac{1}{64\pi^2} \epsilon^{\mu\nu\rho\sigma} \mathrm{tr} F_{\mu\nu} F_{\rho\sigma} \,. \qquad (9.7)$$

In a translationally invariant state, the rate of change of N_{CS} per unit volume V and time t is called the Sphaleron decay rate, denoted by Γ_{CS},

$$\Gamma_{CS} \equiv \frac{\langle (\Delta N_{CS})^2 \rangle}{Vt} = \int d^4x \, \langle q(x^\mu) q(0) \rangle_w = -\lim_{\omega \to 0} \frac{2T}{\omega} G_R(\omega, \mathbf{k} = 0) \,. \qquad (9.8)$$

where the subscript W denotes the Wightman function. The first equality is evident, while the second one stems from the relation between Wightman and retarded Green's functions in a state at equilibrium with temperature T. This quantity determines the likelihood of generating an effective axial chemical potential μ_5 in (9.6), and, eventually the magnitude of CME in heavy ion collisions. QCD at weak coupling yields [40]

$$\Gamma_{CS} = 192.8\alpha_s^5 T^4 , \tag{9.9}$$

where $\alpha_s = g_s^2/(4\pi)$ is the interaction strength.

The question that we want to answer using holography is the magnitude of Γ_{CS} at strong coupling. This question was answered for the maximally supersymmetric $\mathcal{N} = 4$ Yang–Mills theory using AdS/CFT in [41]. The bulk field dual to the CP-odd operator (9.7) is identified with the bulk axion $a(r, x)$ that is a CP-odd massless pseudo-scalar field in the corresponding 5D $\mathcal{N} = 8$ supergravity. This can be computed using the holographic prescription by solving massless bulk field equation of motion with infalling boundary conditions at the horizon and non-normalizable boundary conditions at the boundary. The UV boundary condition is

$$a(r, x) \to \kappa\,\theta, \qquad r \to 0 , \tag{9.10}$$

where θ is the theta-parameter in the QCD Lagrangian $\theta \int d^4x\, \mathrm{tr}\,(F \wedge F)$, because, as mentioned above the bulk axion a couples to the topological charge on the boundary. Here κ is a free parameter which cancels in the end result. Using these boundary conditions, one obtains [41],

$$\Gamma_{CS}\Big|_{conf} = \frac{\lambda^2}{256\pi^3}T^4 , \tag{9.11}$$

where λ is the 't Hooft coupling in the $\mathcal{N} = 4$ super Yang–Mills at large N. CS decay rate is also expected to depend on B, a question that was addressed for $\mathcal{N} = 4$ super Yang–Mills in [42] using the dual black-brane background constructed in [43].

For the QGP, however, we are interested in the analogous results for the strongly interacting, non-conformal plasma, described by the ihQCD model. This calculation was carried out in [2] for vanishing B and in [44] for finite B.

The bulk-axion field can be introduced in the ihQCD model by adding to (7.2) a kinetic term of the form [45, 46]:

$$S_a = M_p^3 \int d^5x \sqrt{-g} Z(\Phi) g^{\mu\nu} \partial_\mu a(x, r) \partial_\nu a(x, r) . \tag{9.12}$$

Note that this term in the action is suppressed as $1/N^2$ compared to the two terms in (7.2) which consistent with the fact that the physics associated with the dual operator (9.7) in QCD is $1/N^2$ suppressed. Practically this means that the field $a(r, x)$ can be treated as a perturbation on top of the ihQCD background obtained in the previous sections.

We included a non-trivial, dilaton dependent kinetic potential $Z(\Phi)$ in (9.12). Its presence in general is expected in compactifications of IIB supergravity to 5D. It is argued to be also present in the effective action of non-critical string theory in [47]. Solving the axion field equation from (9.12) on the black-brane background (6.4) one finds the analytic result:

$$\Gamma_{CS} = \frac{\kappa^2}{N_c^2} \frac{sT}{2\pi} Z(\Phi_h) , \tag{9.13}$$

where s is the entropy density, Φ_h is the value of the dilaton at the horizon, and the constant κ is defined in (9.10). The $\Phi \to -\infty$ limit of $Z(\Phi)$ is fixed by the value of the topological susceptibility $\chi_t = \partial^2 \epsilon(\theta)/\partial\theta^2$ where $\epsilon(\theta)$ is the θ-dependent vacuum energy that is identified with the on-shell action S_a (9.12) in the bulk. This fixes [45]

$$Z \to Z_0 \approx 33.25/\kappa^2, \qquad \Phi \to -\infty , \tag{9.14}$$

where Z_0 is determined by e.g. the lattice data for the topological susceptibility [2, 45]. On the other hand, the IR asymptotics of the function $Z(\Phi)$ can be fixed by "glueball universality" that originates from linear confinement [45], i.e. requiring that the axionic glueball states (excitations of the operator q in (9.7)) carry a mass $m_n^2 \propto n$ in the limit of large excitation number $n \gg 1$. One finds that this asymptotic behavior follows if one requires [45]

$$Z \to c_4 e^{4\Phi} \qquad \Phi \to +\infty , \tag{9.15}$$

where c_4 is a constant. The profile of the function $Z(\Phi)$ for intermediate values of Φ is not completely fixed, but one finds good match with lattice data if one parametrizes this function as

$$Z(\Phi) = Z_0 \left(1 + c_1 e^{\Phi} + c_4 e^{4\Phi} \right) , \tag{9.16}$$

depending on two parameters c_1 and c_4. These constants can then be fixed by matching the lattice data, see e.g. [48]. One finds a large allowed range for these parameters [2]:

$$0 < c_1 < 5, \qquad 0.06 < c_4 < 50 . \tag{9.17}$$

One finds large systematic errors for the physics associated to the CP-odd term (9.12) in ihQCD. We show the result for the sphaleron decay rate Γ_{CS} as a function of temperature in Fig. 9.5. The allowed values for the decay rate is shown by the blue shaded region. The large systematic uncertainty follows from Eq. (9.17) in parametrization of the function $Z(\Phi)$, (9.16). The decay rate shown in the plot is normalized by its value in the limit $\Phi_h \to \infty$ (large T).[4] We observe two important features in Fig. 9.5. First, that it is bounded from below as:

---

[4] The constant κ that appears in this normalization is defined in (9.10).

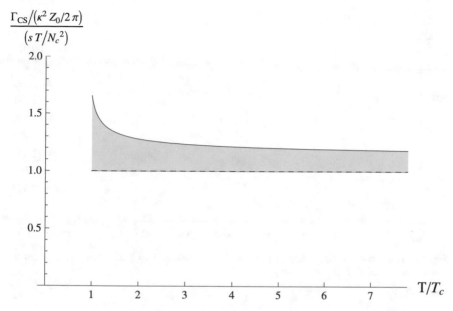

Fig. 9.5 The sphaleron decay rate, properly normalized, as a function of temperature in ihQCD. The blue shaded region are the allowed values for this decay rate, with uncertainties from constants c_1 and c_4 in (9.17). Plot taken from paper [2]

$$\Gamma_{CS}(T) > \frac{\kappa^2}{N_c^2} \frac{s(T)T}{2\pi} Z_0, \qquad T > T_c, \tag{9.18}$$

for all values of T in the deconfined phase. Second, we observe that it is monotonically decreasing with T. This can be shown analytically [2]. This means that sphaleron decays, hence the axial chemical potential μ_5, is more pronounced just above but close to the deconfinement temperature T_c, that is the regime most relevant to the QGP physics. We emphasize that these are *universal features* that follow from ihQCD, regardless of the detailed choices made for the potentials V and Z that enter the ihQCD action.

It is tempting to compare the actual value for the decay rate we obtain from ihQCD for the non-conformal plasma, to the original conformal result (9.11). In the conformal case, for a standard choice $\lambda = 6\pi$ for the QGP, see [49], one finds

$$\left.\frac{\Gamma_{CS}}{T^4}\right|_{conf} \approx 0.045. \tag{9.19}$$

On the other hand, if one calculates (9.13) at T_c using values of parameters quoted above one finds

$$2.8 > \left.\frac{\Gamma_{CS}(T_c)}{T_c^4}\right|_{ihQCD} > 1.64, \tag{9.20}$$

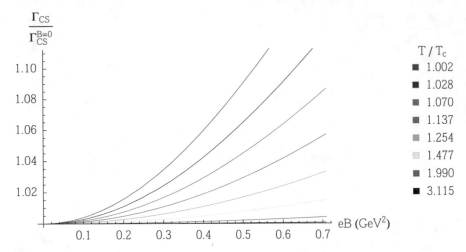

Fig. 9.6 The sphaleron decay rate, normalized by its value at $B = 0$ as a function of B for the various choices of T. Plot taken from paper [44]

which is considerably larger than the conformal value. We conclude ihQCD predicts a stronger presence of sphaleron decays than the conformal plasma modeled by the AdS-Schwarzschild black-brane. Whether this necessitates a larger value of effective axial chemical potential, hence larger possibility for the chiral magnetic effect is unclear.

One can also study dependence of the sphaleron decay rate on magnetic field [44]. For this calculation, one has to use the background with the flavor term that follows from (7.2). The analytic result (9.13) is applicable and yields a function of both B and T through the horizon value of the dilaton, as well as the entropy density s. The resulting profile for the sphaleron decay rate is plotted in Fig. 9.6. We observed that magnetic fields increase the CS decay rate further, near the deconfinement transition.

References

1. R. Kubo, J. Phys. Soc. Jpn. **12**, 570–586 (1957). https://doi.org/10.1143/JPSJ.12.570
2. U. Gürsoy, I. Iatrakis, E. Kiritsis, F. Nitti, A. O'Bannon, JHEP **1302**, 119 (2013). https://doi.org/10.1007/JHEP02(2013)119, arXiv:1212.3894 [hep-th]
3. H.B. Meyer, Phys. Rev. Lett. **100**, 162001 (2008). https://doi.org/10.1103/PhysRevLett.100.162001, arXiv:0710.3717 [hep-lat]
4. S.S. Gubser, A. Nellore, S.S. Pufu, F.D. Rocha, Phys. Rev. Lett. **101**, 131601 (2008). https://doi.org/10.1103/PhysRevLett.101.131601, arXiv:0804.1950 [hep-th]
5. G. Policastro, D.T. Son, A.O. Starinets, Phys. Rev. Lett. **87**, 081601 (2001). https://doi.org/10.1103/PhysRevLett.87.081601, arXiv:hep-th/0104066
6. P. Kovtun, D.T. Son, A.O. Starinets, JHEP **10**, 064 (2003). https://doi.org/10.1088/1126-6708/2003/10/064, arXiv:hep-th/0309213 [hep-th]

7. J. Casalderrey-Solana, H. Liu, D. Mateos, K. Rajagopal, U.A. Wiedemann, https://doi.org/10.1017/CBO9781139136747, arXiv:1101.0618 [hep-th]
8. P. Burikham, N. Poovuttikul, Phys. Rev. D **94**(10), 106001 (2016). https://doi.org/10.1103/PhysRevD.94.106001, arXiv:1601.04624 [hep-th]
9. S.A. Hartnoll, D.M. Ramirez, J.E. Santos, JHEP **03**, 170 (2016). https://doi.org/10.1007/JHEP03(2016)170, arXiv:1601.02757 [hep-th]
10. L. Alberte, M. Baggioli, O. Pujolas, JHEP **07**, 074 (2016). https://doi.org/10.1007/JHEP07(2016)074, arXiv:1601.03384 [hep-th]
11. Y. Ling, Z.Y. Xian, Z. Zhou, JHEP **11**, 007 (2016). https://doi.org/10.1007/JHEP11(2016)007, arXiv:1605.03879 [hep-th]
12. M. Baggioli, W.J. Li, SciPost Phys. **9**(1), 007 (2020). https://doi.org/10.21468/SciPostPhys.9.1.007, arXiv:2005.06482 [hep-th]
13. J. Erdmenger, P. Kerner, H. Zeller, Phys. Lett. B **699**, 301–304 (2011). https://doi.org/10.1016/j.physletb.2011.04.009, arXiv:1011.5912 [hep-th]
14. A. Rebhan, D. Steineder, Phys. Rev. Lett. **108**, 021601 (2012). https://doi.org/10.1103/PhysRevLett.108.021601, arXiv:1110.6825 [hep-th]
15. K.A. Mamo, JHEP **10**, 070 (2012). https://doi.org/10.1007/JHEP10(2012)070, arXiv:1205.1797 [hep-th]
16. S. Jain, N. Kundu, K. Sen, A. Sinha, S.P. Trivedi, JHEP **01**, 005 (2015). https://doi.org/10.1007/JHEP01(2015)005, arXiv:1406.4874 [hep-th]
17. R. Critelli, S.I. Finazzo, M. Zaniboni, J. Noronha, Phys. Rev. D **90**(6), 066006 (2014). https://doi.org/10.1103/PhysRevD.90.066006, arXiv:1406.6019 [hep-th]
18. S. Jain, R. Samanta, S.P. Trivedi, JHEP **10**, 028 (2015). https://doi.org/10.1007/JHEP10(2015)028, arXiv:1506.01899 [hep-th]
19. S.I. Finazzo, R. Critelli, R. Rougemont, J. Noronha, Phys. Rev. D **94**(5), 054020 (2016). [Erratum: Phys. Rev. D **96**(1), 019903 (2017)]. https://doi.org/10.1103/PhysRevD.94.054020, arXiv:1605.06061 [hep-ph]
20. D. Giataganas, U. Gürsoy, J.F. Pedraza, Phys. Rev. Lett. **121**(12), 121601 (2018). https://doi.org/10.1103/PhysRevLett.121.121601, arXiv:1708.05691 [hep-th]
21. U. Gürsoy, M. Järvinen, G. Nijs, J.F. Pedraza, JHEP **03**, 180 (2021). https://doi.org/10.1007/JHEP03(2021)180, arXiv:2011.09474 [hep-th]
22. Q. Li, D.E. Kharzeev, C. Zhang, Y. Huang, I. Pletikosic, A.V. Fedorov, R.D. Zhong, J.A. Schneeloch, G.D. Gu, T. Valla, Nature Phys. **12**, 550–554 (2016). https://doi.org/10.1038/nphys3648, arXiv:1412.6543 [cond-mat.str-el]
23. S.L. Adler, Phys. Rev. **177**, 2426 (1969). https://doi.org/10.1103/PhysRev.177.2426
24. J.S. Bell, R. Jackiw, Nuovo Cim. A **60**, 47 (1969). https://doi.org/10.1007/BF02823296
25. K. Fukushima, D.E. Kharzeev, H.J. Warringa, Phys. Rev. D **78**, 074033 (2008). https://doi.org/10.1103/PhysRevD.78.074033, arXiv:0808.3382 [hep-ph]
26. D.E. Kharzeev, K. Landsteiner, A. Schmitt, H.U. Yee, Lect. Notes Phys. **871**, 1 (2013). arXiv:1211.6245 [hep-ph]
27. D. Kharzeev, K. Landsteiner, A. Schmitt, H.U. Yee, Lect. Notes Phys. **871**, 1 (2013). https://doi.org/10.1007/978-3-642-37305-3
28. V.A. Miransky, I.A. Shovkovy, Phys. Rept. **576**, 1 (2015). https://doi.org/10.1016/j.physrep.2015.02.003, arXiv:1503.00732 [hep-ph]
29. V.I. Zakharov, Lect. Notes Phys. **871**, 295 (2013). arXiv:1210.2186 [hep-ph]
30. K. Landsteiner, E. Megias, F. Pena-Benitez, Phys. Rev. Lett. **107**, 021601 (2011). https://doi.org/10.1103/PhysRevLett.107.021601, arXiv:1103.5006 [hep-ph]
31. K. Landsteiner, arXiv:1610.04413 [hep-th]
32. A. Vilenkin, Phys. Rev. Lett. **41**, 1575 (1978). https://doi.org/10.1103/PhysRevLett.41.1575
33. D.E. Kharzeev, L.D. McLerran, H.J. Warringa, Nucl. Phys. A **803**, 227 (2008). https://doi.org/10.1016/j.nuclphysa.2008.02.298, arXiv:0711.0950 [hep-ph]
34. M.E. Peskin, D.V. Schroeder,
35. D.T. Son, P. Surowka, Phys. Rev. Lett. **103**, 191601 (2009). https://doi.org/10.1103/PhysRevLett.103.191601, arXiv:0906.5044 [hep-th]

36. N.S. Manton, Phys. Rev. D **28**, 2019 (1983). https://doi.org/10.1103/PhysRevD.28.2019
37. V.A. Kuzmin, V.A. Rubakov, M.E. Shaposhnikov, Phys. Lett. **155B**, 36 (1985). https://doi.org/10.1016/0370-2693(85)91028-7
38. P.B. Arnold, L.D. McLerran, Phys. Rev. D **36**, 581 (1987). https://doi.org/10.1103/PhysRevD.36.581
39. P.B. Arnold, L.D. McLerran, Phys. Rev. D **37**, 1020 (1988). https://doi.org/10.1103/PhysRevD.37.1020
40. P.B. Arnold, D. Son, L.G. Yaffe, Phys. Rev. D **55**, 6264 (1997). https://doi.org/10.1103/PhysRevD.55.6264, arXiv:hep-ph/9609481
41. D.T. Son, A.O. Starinets, JHEP **0209**, 042 (2002). https://doi.org/10.1088/1126-6708/2002/09/042, arXiv:hep-th/0205051
42. G. Basar, D.E. Kharzeev, Phys. Rev. D **85**, 086012 (2012). https://doi.org/10.1103/PhysRevD.85.086012, arXiv:1202.2161 [hep-th]
43. E. D'Hoker, P. Kraus, JHEP **0910**, 088 (2009). https://doi.org/10.1088/1126-6708/2009/10/088, arXiv:0908.3875 [hep-th]
44. T. Drwenski, U. Gursoy, I. Iatrakis, arXiv:1506.01350 [hep-th]
45. U. Gursoy, E. Kiritsis, F. Nitti, JHEP **0802**, 019 (2008). https://doi.org/10.1088/1126-6708/2008/02/019, arXiv:0707.1349 [hep-th]
46. U. Gursoy, E. Kiritsis, L. Mazzanti, F. Nitti, JHEP **0905**, 033 (2009). https://doi.org/10.1088/1126-6708/2009/05/033, arXiv:0812.0792 [hep-th]
47. U. Gursoy, E. Kiritsis, JHEP **0802**, 032 (2008). https://doi.org/10.1088/1126-6708/2008/02/032, arXiv:0707.1324 [hep-th]
48. C.J. Morningstar, M.J. Peardon, Phys. Rev. D **60**, 034509 (1999). https://doi.org/10.1103/PhysRevD.60.034509, arXiv:hep-lat/9901004
49. S.S. Gubser, Phys. Rev. D **76**, 126003 (2007). https://doi.org/10.1103/PhysRevD.76.126003, arXiv:hep-th/0611272

Chapter 10
Conclusion and a Look Ahead

These notes aim at a self-contained introduction to applications of the gauge/gravity correspondence to the theory of strong interactions, QCD, with an emphasis on describing the thermodynamic and transport properties of the quark-gluon plasma. We outlined construction of the improved holographic QCD model, explained how to fix its parameters, the resulting structure of the ground and thermal states, calculation of thermodynamic observables and their comparison to the first-principles theory i.e. lattice QCD, how the holographic and hydrodynamic descriptions are merged and the calculation of the transport coefficients such as the bulk and the shear viscosities. We also focused on the electromagnetic properties of strong interactions within the holographic approach to QCD. In particular, we discussed what holography has to say about some of the open problems such as how the magnetic fields affect the phase diagram, how do the transport coefficients depend on it, and the underlying mechanisms behind recently discovered phenomena such as the inverse magnetic catalysis.

This bottom-up holographic approach to QCD successfully captures the salient features of QCD in the IR regime and matches well the thermodynamic observables and the hadron spectra calculated on the lattice. It then extends it to the problems that cannot be easily accessed by the lattice simulations: finite quark density and transport. Even though the last word should be said by the first-principles theory, the holographic approach provides valuable insights into these problems capturing the correct qualitative behavior. Moreover, it provides conclusive answers for the particular case of maximally supersymmetric Yang-Mills theory at large N and infinite coupling.

Among such qualitative insights are: a direct connection between linear confinement of quarks and discrete and gapped hadron spectrum, presence of a deconfinement temperature at finite T for any holographic gauge theory that exhibits linear confinement in the ground state, a universal increase in transport coefficients such

U. Gürsoy, *Holography and Magnetically Induced Phenomena in QCD*, SpringerBriefs in Physics, https://doi.org/10.1007/978-3-030-79599-3_10

as the bulk viscosity and the sphaleron decay rate as temperature approaches the deconfinement temperature from above, and universal bounds on diffusion constants that describe the energy-momentum loss of probe quarks in the plasma. One should stress that, even if qualitative in flavor, these are definite results that remain inaccessible by the traditional methods of quantum field theory to date. They become easy exercises of real analysis and geometry in the holographic description.[1]

On top of these qualitative insights, quantitative predictions of the specific holographic model we present in these notes, i.e. the improved holographic QCD, are also quite, perhaps unexpectedly, good. It would be very desirable to be able to place systematic errors on the holographic description of the quark-gluon plasma, which would promote it to a standard technique in analyzing experimental data, but this seems hard to achieve today.

Among the various various important topics which we deliberately left out in this monograph, the following stand out:

- Non-equilibrium thermalization process in strongly coupled non-conformal gauge theories, in particular QCD. Underpinning the precise mechanism(s) behind the rapid thermalization of quarks and gluons produced in the heavy ion collisions is a long-standing open problem. The study of holographic thermalization has started with the pioneering work of Chesler and Yaffe [1]. This work and most of the subsequent developments focused on thermalization and out-of equilibrium physics in conformal rather than non-conformal plasmas. The latter only started attracting attention recently with the works [2–8]. We have not reviewed these developments in this review due to lack of space.

- Description of baryons in holographic QCD. Fluctuations of the bulk fields in the 5D background yield the holographic prediction for the meson and glueball spectra, but not the baryon spectrum. Baryons in holography should be treated separately as soliton configurations as was first demonstrated in [9, 10]. Baryons in the context of simple bottom-up holographic models was first discussed in [11, 12]. In ihQCD they are first addressed in detail in the recent work of [13]. These works introduce a coarse-grained description of the baryons in holography leaving ample room for further development.

- Anomalous transport at strong coupling. Radiative and non-perturbative corrections to chiral magnetic and chiral vortical conductivities in the presence of dynamical gluon or photon field to the axial anomaly is an open problem which has been addressed using holographic methods [14, 15]. These works only scratched the surface of this notorious problem which would clearly benefit from further holographic studies.

The quest for developing a realistic holographic model for QCD continues to be an exciting scientific journey that has already provided us with an analytic tool to elucidate problems that haunted high energy physicists for long. It seems it will

---

[1] Another striking example from a different application of holography is the holographic proof of strong subadditivity of the entanglement entropy; an extremely hard problem in quantum field theory, reduced to a very simple geometric exercise in holography.

continue to fuel theoretical advances in study of the quark-gluon plasma, and other strongly interacting plasmas akin to it, by providing an overarching approach where the many different facets of the physics ranging from confinement to non-equilibrium physics can be treated under the single umbrella, a unique 5D gravitational action.

References

1. P.M. Chesler, L.G. Yaffe, Phys. Rev. Lett. **102**, 211601 (2009). https://doi.org/10.1103/PhysRevLett.102.211601 [arXiv:0812.2053] [hep-th]
2. T. Ishii, E. Kiritsis, C. Rosen, JHEP **08**, 008 (2015). https://doi.org/10.1007/JHEP08(2015)008 [arXiv:1503.07766] [hep-th]
3. A. Buchel, M.P. Heller, R.C. Myers, Phys. Rev. Lett. **114**(25), 251601 (2015). https://doi.org/10.1103/PhysRevLett.114.251601 [arXiv:1503.07114] [hep-th]
4. J.F. Fuini, L.G. Yaffe, JHEP **07**, 116 (2015). https://doi.org/10.1007/JHEP07(2015)116 [arXiv:1503.07148] [hep-th]
5. R.A. Janik, G. Plewa, H. Soltanpanahi, M. Spalinski, Phys. Rev. D **91**(12), 126013 (2015). https://doi.org/10.1103/PhysRevD.91.126013 [arXiv:1503.07149] [hep-th]
6. R.A. Janik, J. Jankowski, H. Soltanpanahi, JHEP **06**, 047 (2016). https://doi.org/10.1007/JHEP06(2016)047 [arXiv:1603.05950] [hep-th]
7. U. Gursoy, M. Jarvinen, G. Policastro, JHEP **01**, 134 (2016). https://doi.org/10.1007/JHEP01(2016)134 [arXiv:1507.08628] [hep-th]
8. M. Attems, J. Casalderrey-Solana, D. Mateos, I. Papadimitriou, D. Santos-Oliván, C.F. Sopuerta, M. Triana, M. Zilhão, JHEP **10**, 155 (2016). https://doi.org/10.1007/JHEP10(2016)155 [arXiv:1603.01254] [hep-th]
9. E. Witten, JHEP **07**, 006 (1998). https://doi.org/10.1088/1126-6708/1998/07/006 [arXiv:hep-th/9805112] [hep-th]
10. D.J. Gross, H. Ooguri, Phys. Rev. D **58**, 106002 (1998). https://doi.org/10.1103/PhysRevD.58.106002 [arXiv:hep-th/9805129] [hep-th]
11. A. Pomarol, A. Wulzer, Nucl. Phys. B **809**, 347–361 (2009). https://doi.org/10.1016/j.nuclphysb.2008.10.004 [arXiv:0807.0316] [hep-ph]
12. G. Panico, A. Wulzer, Nucl. Phys. A **825**, 91–114 (2009). https://doi.org/10.1016/j.nuclphysa.2009.04.004 [arXiv:0811.2211] [hep-ph]
13. T. Ishii, M. Järvinen, G. Nijs, JHEP **07**, 003 (2019). https://doi.org/10.1007/JHEP07(2019)003 [arXiv:1903.06169] [hep-ph]
14. U. Gürsoy, A. Jansen, JHEP **10**, 092 (2014). https://doi.org/10.1007/JHEP10(2014)092 [arXiv:1407.3282] [hep-th]
15. A. Jimenez-Alba, K. Landsteiner, L. Melgar, Phys. Rev. D **90**, 126004 (2014). https://doi.org/10.1103/PhysRevD.90.126004 [arXiv:1407.8162] [hep-th]

Appendices

A.1 Scalar Variables

First rewrite the Einstein's equations for the black-brane background in the domain-wall coordinate system

$$ds^2 = e^{2A(u)} \left(\frac{du^2}{f(u)} + \delta_{ij} dx^i dx^j \right) - f(u) dt^2 , \qquad (A.1)$$

that is related to (the Lorentzian version of) (6.4) by a coordinate transformation $du = \exp(Ar)dr$. The Einstein's equations (6.9) in this coordinate system read:

$$A'' = -\frac{4}{9}(\Phi')^2, \qquad 3A'' + 12(A')^2 + 3A'\frac{f'}{f} = \frac{e^{2A}}{f} V(\Phi), \qquad f'' + 4A'f' = 0 . \qquad (A.2)$$

Now define

$$D(\Phi) \equiv A' , \qquad (A.3)$$

and use the chain rule for derivatives to solve the first equation in (A.2) for D:

$$D(\Phi) = -\frac{1}{\ell} e^{-\frac{4}{3} \int_0^\Phi d\Phi X(\Phi)} , \qquad (A.4)$$

where we used the definition (6.10). Then again using the chain rule in the third equation in (A.2) and taking into account the definitions on obtains the equation of motion for the Y scalar variable (6.12). Now, use the chain rule to rewrite the second equation in (A.2) in terms of X, Y, D and their derivatives, take the logarithmic derivative of this equation with respect to Φ, use the solution (A.4) and the Eq. (6.12) derived above and simplify to obtain the equation of motion for the X scalar variable, Eq. (6.11). All in all we derived

© The Author(s), under exclusive license to Springer Nature Switzerland AG 2021
U. Gürsoy, *Holography and Magnetically Induced Phenomena in QCD*,
SpringerBriefs in Physics,
https://doi.org/10.1007/978-3-030-79599-3

$$\frac{dX}{d\Phi} = -\frac{4}{3}(1 - X^2 + Y)\left(1 + \frac{3}{8}\frac{1}{X}\frac{d\log V}{d\Phi}\right), \tag{A.5}$$

$$\frac{dY}{d\Phi} = -\frac{4}{3}(1 - X^2 + Y)\frac{Y}{X}. \tag{A.6}$$

These are two first order equations. The total degree of the system of Einstein's equations is 5. The rest of the equations follow from (A.4) and the definitions (6.10) as

$$A' = -\frac{1}{\ell}e^{-\frac{4}{3}\int_0^\Phi d\Phi X(\Phi)} \tag{A.7}$$

$$\Phi' = -\frac{3X}{\ell}e^{-\frac{4}{3}\int_0^\Phi d\Phi X(\Phi)} \tag{A.8}$$

$$g' = -\frac{4Y}{\ell}e^{-\frac{4}{3}\int_0^\Phi d\Phi X(\Phi)} \tag{A.9}$$

where we defined $g = \log f$. These equations complete the system. The corresponding equations for the thermal gas solution (6.3) can be obtained from these by setting $Y = 0$.

A.2   The Potentials

In this appendix we list the potentials of the V-QCD model. We define $\lambda = \exp\Phi$. The potentials read:

$$V(\lambda) = \frac{12}{\mathcal{L}_0^2}\left[1 + \frac{88\lambda}{27} + \frac{4619\lambda^2}{729}\frac{\sqrt{1 + \ln(1 + \lambda)}}{(1 + \lambda)^{2/3}}\right], \tag{A.10}$$

$$V_{f0} = \frac{12}{\mathcal{L}_{UV}^2}\left[\frac{\mathcal{L}_{UV}^2}{\mathcal{L}_0^2} - 1 + \frac{8}{27}\left(11\frac{\mathcal{L}_{UV}^2}{\mathcal{L}_0^2} - 11 + 2x\right)\lambda \right.$$
$$\left. + \frac{1}{729}\left(4619\frac{\mathcal{L}_{UV}^2}{\mathcal{L}_0^2} - 4619 + 1714x - 92x^2\right)\lambda^2\right],$$

$$\kappa(\lambda) = \frac{[1 + \ln(1 + \lambda)]^{-1/2}}{\left[1 + \frac{3}{4}\left(\frac{115 - 16x}{27} - \frac{1}{2}\right)\lambda\right]^{4/3}}, \qquad a(\lambda) = \frac{3}{2\mathcal{L}_{UV}^2}, \tag{A.11}$$

where $\mathcal{L}_{UV}$ is the AdS radius, so that the boundary expansion of the metric is $A \sim \ln(\mathcal{L}_{UV}/r) + \cdots$. The radius depends on x as

$$\mathcal{L}_{UV}^3 = \mathcal{L}_0^3\left(1 + \frac{7x}{4}\right). \tag{A.12}$$

The function w is parametrized by a single parameter c

$$w(\lambda) = \kappa(c\lambda) = \frac{(1 + \log(1 + c\lambda))^{-\frac{1}{2}}}{\left(1 + \frac{3}{4}\left(\frac{115-16x}{27} - \frac{1}{2}\right)c\lambda\right)^{4/3}}, \tag{A.13}$$

where x is the ratio of the number of flavors to color.

A.3 Equations of Motion

The Einstein equations of motion from (5.1) and (8.1) read

$$R_{\mu\nu} - \frac{1}{2}g_{\mu\nu}R - \left(\frac{4}{3}\partial_\mu\Phi\partial_\nu\Phi - \frac{2}{3}(\partial\Phi)^2 g_{\mu\nu} + \frac{1}{2}g_{\mu\nu}V(\Phi)\right)$$
$$-x\frac{V_f(\Phi,\tau)}{2}\left(-g_{\mu\nu}\sqrt{D} + \frac{1}{\sqrt{D}}\frac{dD}{dg^{\mu\nu}}\right) = 0, \tag{A.14}$$

where $D = det\left(\delta^\mu_\lambda + w(\Phi)\, g^{\mu\nu}\, V_{\nu\lambda} + \kappa(\Phi)g^{\mu\nu}\partial_\nu\tau\,\partial_\lambda\tau\right)$. Inserting here our Ansätze for the metric and the gauge field gives

$$3\,A'' + \frac{2}{3}\Phi'^2 + 3\,A'^2 + (3\,A' - W')\frac{f'}{2f} + \frac{x\,V_f(\Phi,\tau)\,G\,e^{2A}}{2\,Q\,f}(2\,Q^2 - 1) - \frac{e^{2A}}{2f}V(\Phi) = 0, \tag{A.15}$$

$$W'' + \frac{W'f'}{f} + W'^2 + 3\,A'W' + \frac{x\,V_f(\Phi,\tau)\,G\,e^{2A}}{2\,Q\,f}\left(1 - Q^2\right) = 0, \tag{A.16}$$

$$f'' + (3\,A' + W')\,f' + \frac{x\,V_f(\Phi,\tau)\,e^{2A}\,G}{Q}\left(1 - Q^2\right) = 0, \tag{A.17}$$

where we defined

$$G(r) = \sqrt{1 + e^{-2A(r)}\kappa(\Phi,\tau)f(r)(\tau\partial_r\tau)^2},$$
$$Q(r) = \sqrt{1 + w(\Phi)^2 B^2 e^{-4A(r)}}. \tag{A.18}$$

The first order constraint equation reads

$$\frac{2}{3}\Phi'^2 - (3\,A' + W')\frac{f'}{2f} - 6\,A'^2 - 3\,A'\,W' + \frac{e^{2A}}{2f}V(\Phi) - \frac{x\,V_f(\Phi,\tau)\,Q\,e^{2A}}{2\,G\,f} = 0. \tag{A.19}$$

The dilaton equation of motion becomes

$$\Phi'' + \left(3\,A' + W' + \frac{f'}{f}\right)\Phi' + \frac{3}{8}\frac{e^{2A}}{f}\,\partial_\Phi V(\Phi) - \frac{3\,x\,B^2\,e^{-2A}\,G\,V_f(\Phi,\tau)w(\Phi)}{8\,f\,Q}\partial_\Phi w(\Phi)$$

$$-\frac{3\,x\,e^{2A}\,G\,Q}{8\,f}\partial_\Phi V_f(\Phi,\tau) - \frac{3\,x\,Q\,V_f(\Phi,\tau)\,\tau'^2}{16\,G}\partial_\Phi\kappa(\Phi) = 0, \tag{A.20}$$

and the tachyon equation of motion is

$$\tau'' - \frac{e^{2A}\,G^2}{f\,\kappa(\Phi)}\partial_\tau\,\log\,V_f(\Phi,\tau) + e^{-2A}\,f\,\kappa(\Phi)\left(W' + \frac{1}{2}\frac{f'}{f}\right.$$

$$+ 2\,A'\frac{1+Q^2}{Q^2} + \frac{1}{2}\,\Phi'\,\partial_\Phi\,\log\,(\kappa(\Phi)\,V_f(\Phi,\tau)^2) - \frac{\Phi'\,(1-Q^2)}{Q^2}\partial_\Phi\,\log\,w(\Phi)\bigg)\tau'^3$$

$$+ \left(A'\frac{2+Q^2}{Q^2} + W' + \frac{f'}{f} + \Phi'\partial_\Phi\,\log(V_f(\Phi,\tau)\,\kappa(\Phi)) - \frac{\Phi'\,(1-Q^2)}{Q^2}\partial_\Phi\,\log\,w(\Phi)\right)\tau' = 0. \tag{A.21}$$

Printed in the United States
by Baker & Taylor Publisher Services